Routledge Revivals

Public Expenditure Decisions in the Urban Community

In 1962, the Committee on Urban Economics held a conference on public expenditure decisions in order to promote analysis of the issues facing the public sector of the urban economy. Originally published in 1965, this report pulls together key papers presented at this conference discussing issues such as urban services, the patterns of public expenditure and the quality of government services in urban areas to draw conclusions on the difficulties of analysis and how economic tools could be utilised more effectively to solve these difficulties. This title will be of interest to students of environmental studies and economics.

Public Expenditure Decisions in the Urban Community

Edited by
Howard G. Schaller

First published in 1965
by Resources for the Future, Inc.

This edition first published in 2016 by Routledge
2 Park Square, Milton Park, Abingdon, Oxon, OX14 4RN
and by Routledge
711 Third Avenue, New York, NY 10017

Routledge is an imprint of the Taylor & Francis Group, an informa business

Publisher's Note
The publisher has gone to great lengths to ensure the quality of this reprint but points out that some imperfections in the original copies may be apparent.

Disclaimer
The publisher has made every effort to trace copyright holders and welcomes correspondence from those they have been unable to contact.

A Library of Congress record exists under LC control number: 63022774

ISBN 13: 978-1-138-12027-3 (hbk)
ISBN 13: 978-1-31565-175-0 (ebk)
ISBN 13: 978-1-138-12030-3 (pbk)

PUBLIC EXPENDITURE DECISIONS IN THE URBAN COMMUNITY

PUBLIC EXPENDITURE

IN THE

Distributed by
The Johns Hopkins Press, Baltimore and London

DECISIONS

URBAN COMMUNITY

Papers presented at a conference, May 14-15, 1962, under the sponsorship of the Committee on Urban Economics of Resources for the Future, Inc.

Edited by
Howard G. Schaller

Resources for the Future, Inc.
1755 Massachusetts Avenue, N.W., Washington, D.C. 20036

ISBN 0-8018-0576-7

Originally published, 1963
Second printing, 1965
Third printing, 1968
Fourth printing, 1971

Resources for the Future is a nonprofit corporation for research and education in the development, conservation, and use of natural resources and the improvement of the quality of the environment. It was established in 1952 with the cooperation of the Ford Foundation. Part of the work of Resources for the Future is carried out by its resident staff; part is supported by grants to universities and other nonprofit organizations. Unless otherwise stated, interpretations and conclusions in RFF publications are those of the authors; the organization takes responsibility for the selection of significant subjects for study, the competence of the researchers, and their freedom of inquiry.

The conference at which these papers were presented was part of the program of RFF's Committee on Urban Economics. Members of the Committee are: Harvey S. Perloff, of RFF (chairman); Harold J. Barnett, Washington University; Joseph L. Fisher, of RFF; Lyle C. Fitch, Institute of Public Administration; Alvin H. Hansen, formerly of Harvard University; Walter W. Heller, Council of Economic Advisers; Werner Z. Hirsch, University of California (Los Angeles); Edgar M. Hoover, University of Pittsburgh; Howard G. Schaller, Tulane University; Leo F. Schnore, University of Wisconsin; Arthur M. Weimer, Indiana University; and Robert C. Wood, Massachusetts Institute of Technology.

RFF editors: Henry Jarrett, Vera W. Dodds, Nora E. Roots, Tadd Fisher.

Preface

As urban communities have become larger and more complex and their problems more pressing, an increasing number of scholars from various disciplines have turned attention to them. Yet economists have hesitated to enter this particular intellectual fray, in spite of the fact that our economy is now so largely urban in character. To stimulate interest and advance education and research on the urban economy, a Committee on Urban Economics was organized several years ago by Resources for the Future, composed of economists and other social scientists who had become intrigued by the intellectual and practical challenges of the field. The Committee initiated and has provided continuing guidance for a broad program of fellowships, conferences, research grants, commissioned papers, and other activities, carried out under a special grant to RFF from The Ford Foundation.

The Committee felt that the public sector of the urban economy needed particular attention. The complex of governmental decisions on the urban scene seemed to provide a fruitful framework for advancing the urgently needed evolution of a "Public Economics" of the city. In addition to the traditional task of providing (and financing) a wide range of collective services, such as education and police, local governments have been given the responsibility for guiding the development of urban communities, through city planning and zoning, through construction of transportation facilities and urban renewal and through other long-term capital investments. How best could these service and developmental expenditure decisions be made?

To determine what economics had to say on this subject and what might be the more productive lines of further inquiry, the Committee invited a number of scholars who had in the past done significant work on the problems of public decisions to prepare papers on the special issues arising in the public sector of the urban economy. A conference on Public Expenditure Decisions in the Urban Community, which was centered on these papers, was held May 14-15, 1962, at Resources for the Future in Washington. An unusually able group of discussants, in a series of probing reviews of the papers, presented materials which greatly helped the authors to sharpen and enrich their initial manuscripts.

Howard Schaller, himself a member of the Committee on Urban Economics, undertook the difficult task of editing the papers. The Committee owes a great debt to him.

An important outcome of the conference was the formation of a Committee on Urban Public Expenditures composed of the following members: Jesse Burkhead (Syracuse University); Anthony Downs (Real Estate

Research Corporation, Chicago, Illinois); John Dyckman (Arthur D. Little Co., San Francisco, California); Lyle Fitch (Institute of Public Administration, New York, New York); Julius Margolis, (University of California, Berkeley) chairman of the committee; Jerome W. Milliman (Indiana University); Dick Netzer (New York University); Jerome Rothenberg (Northwestern University); and Howard Schaller (Tulane University).

The committee will sponsor additional conferences and seek in other ways to encourage communication among scholars and practitioners concerned with improving approaches to public decision making in urban communities.

Harvey S. Perloff

Editor's Introduction

The Committee on Urban Economics organized the Conference on Public Expenditure Decisions in the Urban Community in an effort to assess the state of the art of applying the tools of economic analysis to the problems of the public sector of the urban economy. The Conference was held in Washington on May 14-15, 1962. Following the Conference, the Committee decided that publication of the papers might prove helpful to future research in the field. It is hoped that the papers will prove useful not only to economists but to planners, political scientists, sociologists, and others interested in the urban field.

Each of the papers in the volume can be read independently, but their order of appearance may be helpful to readers who intend to pursue the entire volume. The first three papers are more general in nature. William Baumol's paper examines the need for the public production of goods and services and how recent changes in the private sector have sharply affected the need for expansion of the public sector of the urban economy. Allen Manvel examines the changing pattern of local government expenditures and indicates changes that may be forthcoming in the future. Selma Mushkin analyzes the effect of intergovernmental fiscal aids on the local decision-making process and suggests how such aids may be used for improving the process.

The next four papers deal more specifically with the application of theoretical tools to decision-making in the public sector of the urban economy. William Vickrey's paper contains some interesting suggestions for the extension of marginal-cost pricing to the public sector. Russell Ackoff's paper assesses the possibility of using the decision-models developed by operations researchers as an aid to local government decision making. Julius Margolis and Nathaniel Lichfield examine the possibility of more extensive use of benefit-cost analysis in making local government decisions. The last of the four, the paper by Roland McKean, considers the problems that emerge in attempting to apply rational decision models when there is a divergence between individual interest and total costs and gains.

The last two papers deal with problems of measurement. Werner Hirsch considers the vexing question of how to measure the quality of local government output; a question which needs attention if research in the field is to progress. Seymour Sacks suggests that the customary measurements using population density and per capita expenditures can be profitably supplemented in many cases by the use of property valuation per square mile and expenditures per mile measures.

While the Conference did not have a set theme, the papers do underscore a few points. First, they emphasize the great (and familiar) difficulties involved for analysis and policy prescription when resource use is not directed by the market mechanism. Secondly, the papers indicate a growing but as yet meager effort on the part of economists to analyze the problem of collective decisions. As Tibor Scitovsky stated at the Conference,

> More and more we realize the increasing importance of public expenditure on collective goods; and yet most economists keep their noses glued to the problem of how consumers' choice influences (or could best influence) producers' decisions through the market mechanism; we [economists] have failed dismally and scandalously not only to answer but even merely to ask the question of how best to make collective decisions relating to public expenditures on collective goods, so that these decisions will be in closest conformity to the public's preferences.

Finally, the papers indicate that even given the difficulties of the problems and the paucity of efforts in the past, there is hope for the future. Several papers have interesting suggestions as to how the tools of economic analysis might be sharpened and used more extensively.

I wish to thank the following discussants at the Conference for sending me their written comments for use by the authors and in the editorial process: Dick Netzer, F. M. Labovitz, Alice Rivilin, Tibor Scitovsky, Jerome Rothenberg, John Dyckman, Wilbur Thompson, Richard Musgrave, Jerome Milliman, and John Krutilla.

I also wish to thank Harvey Perloff and Lowdon Wingo, Jr., of the Committee on Urban Economics and Resources for the Future, Inc., for their assistance.

HOWARD G. SCHALLER
Tulane University

Contents

PREFACE, Harvey S. Perloff, Resources for the Future, Inc. iii

EDITOR'S INTRODUCTION, Howard G. Schaller, Tulane University of Lousiana v

1. Urban Services: Interactions of Public and Private Decisions, by William J. Baumol, Princeton University 1

2. Changing Patterns of Local Urban Expenditure, by Allen D. Manvel, Bureau of the Census, U.S. Department of Commerce . . 19

3. Intergovernmental Aspects of Local Expenditure Decisions, by Selma J. Mushkin, Advisory Commission on Intergovernmental Relations 37

4. General and Specific Financing of Urban Services, by William W. Vickrey, Columbia University 62

5. Toward Quantitative Evaluation of Urban Services, by Russell L. Ackoff, Case Institute of Technology 91

6. Benefit-Cost Analysis as a Tool in Urban Government Decision Making, by Nathaniel Lichfield, University College of London, and Julius Margolis, University of California (Berkeley) 118

7. Costs and Benefits from Different Viewpoints, by Roland N. McKean, The RAND Corporation 147

8. Quality of Government Services, by Werner Z. Hirsch, University of California (Los Angeles) 163

9. Spatial and Locational Aspects of Local Government Expenditures, by Seymour Sacks, Syracuse University 180

1

Urban Services: Interactions of Public and Private Decisions

by William J. Baumol

As we grope toward more satisfactory criteria for urban public service decision making it is essential to keep in mind the relationship between the operations of the public and the private sectors. Indeed, our political philosophy leads us to consider the government to be, in some ultimate sense, a means whereby the members of the community can have their needs and wants satisfied more effectively. That is, public outlays are only justifiable in terms of their effects on private decisions and activities and, thus, on the welfare of the public in general. Looked at the other way, this means that only changes in the private arrangements can render appropriate types of urban governmental services which were formerly uncalled for. The entire relationship may even be viewed as a dynamic process whereby changes in public services affect private activities, which in turn lead to further changes in public policies and so on ad infinitum.

In this paper I investigate what can be said about the subject both in terms of general principles and by way of illustration in terms of concrete cases. The analysis is centered around three major questions--

1. What advantages can be hoped for by the private sector from the substitution of public for private urban services?
2. How should changing conditions in the private sector affect the provision of public services?
3. What differences result from alternative methods of public participation in the provision of services?

THE RATIONALE OF PUBLIC SERVICES IN A FREE ENTERPRISE SYSTEM

Behind the entire discussion there must lie the basic principles involved in the provision of services by the public sector. We must understand precisely why it is considered appropriate in a free enterprise system to have the government intervene at all. The answer to our question is by no means trivial. After all, a widely held premise is the idea that, other things being equal, the extension of governmental activity is undesirable *per se*. Underlying this assertion there exist more subtle arguments which question the desirability of intervention. For example, there is the view that if the public desires any service sufficiently it will be willing to pay the requisite price. That is, on this assertion, any service supplied or subsidized by the government is likely to represent a misallocation of resources because it involves the provision of a good which is not worth enough to consumers for them to be willing to pay its cost. Or, in more general terms, it is sometimes argued that every governmental decision substitutes fiat for freedom of choice by the public, and any such decision must therefore be inferior, from the point of view of those affected, to what they would have decided for themselves.

These arguments are unlikely to appeal to many of us emotionally. We feel, somehow, that the market mechanism cannot be left to run itself entirely. It needs at least an occasional oiling, some maintenance, and a guard to see to it that no one sabotages the machinery. Even the most extreme of laissez-faire economists have admitted this. But what determines precisely whether a particular governmental service is desirable? Indeed, how can it be that a government, whose stated purpose is to serve its citizens, can improve their welfare by making some decisions for them?

Several grounds for governmental activity have been enunciated in the literature:

Ignornance or incompetence on the part of the public. Essentially, this is the basis on which governments intervene to protect minors or the mentally retarded. It is felt that such people simply do not know enough to take care of themselves and so someone must step in to protect their interests. These grounds are not likely to appeal to us in their most blatant form because we prefer to flatter ourselves on our knowledge and mental competence. Yet this argument provides the justification for a considerable variety of governmental measures. Examples are the pure food and drug act, which requires disclosure of information about ingredients, Securities and Exchange Commission regulations, which insist on the supply of data relating to securities issued on an interstate basis, and a variety of information services supplied by the Department of Agriculture and other government agencies. However, this rationale for government activity does not seem to play a very considerable role in the supply of urban public services and so we will not have occasion to return to it in this paper.

Prevention or amelioration of unfair arrangements. A second ground for government intervention is the feeling that the order of things involves basic inequities which should somehow be mitigated or at least prevented from getting out of hand. Police protection of the weak or honest from the stronger or unscrupulous falls under this head. Another major subject for concern here is poverty and inequality of opportunity resulting from disparities in the distribution of wealth and income. This second element in the rationale of governmental endeavor can be very significant for urban services. Free public education, slum clearance, and police activities are all substantial items in a typical metropolitan budget, and clearly they all relate to the protection of the individual who is either underprivileged or unprepared to defend himself against unlawful elements in the community.

Interdependence of members of the public. There is a third and far more subtle ground on which government activity can be justified in a free society. This is the view that the welfare of any member of the public does not depend on himself alone. Whatever anyone else does is likely to affect him and vice versa. As a result, given two self-interested individuals, A and B, if each independently goes about making the decisions which are best for him alone, the result may well be that the welfare of each falls well below the maximum available to him. A will have acted without taking into account the consequences of his behavior for B, and vice versa. As a result each will suffer from the want of consideration on the part of the other and hence they will both lose. In such a situation they may both benefit if they can agree on a third party (the government) who will see to it that they simultaneously act in a manner calculated to maximize their joint welfare, however defined.

This argument must not be understood to state that selfishness provides its own penalties. Neither individual acting by himself is likely to have the ability to produce a communal optimum by means of his own unsupported activity. Only governmental intervention can in this case offer the individual the opportunity to maximize his welfare, and to achieve what he really wishes. A concrete example may make this clear. Consider a street which is badly paved and which is adding to the automobile repair expenses of its residents (as well as to the costs incurred by motorists in transit). Suppose repaving can be financed and amortized at an average annual cost of ten dollars per resident and that this will (on the average) save him twenty-five dollars per year in automotive maintenance. Clearly, it is in everyone's interest to have the street repaired.

But no one individual can afford to do the repaving by himself. Moreover, if the hat were passed about the neighborhood each resident might well feel that his own individual contribution will not make the difference between failure and success of the enterprise. If, as a result, enough people hold out, the street is likely to retain its decrepit state, to everyone's loss. However, these same citizens who hesitate to contribute on a voluntary basis may reasonably support a special tax bill which requires each and every resident to pay his ten-dollar share of the cost of the improvement, if he knows that in this way he may expect to come out fifteen dollars ahead.

This rather subtle argument covers a wide variety of governmental activities. Familiar examples are the prevention of air pollution, the provision of public parks, street lighting, etc. Other, perhaps more timely illustrations will appear later in this paper.

The argument may conveniently be recapitulated as follows: The activities of and provision for the individual are likely to have so-called external effects--what happens to or is done by A, has a bearing for the welfare of B. These external effects are in many situations likely to be so widely diffused and spread about in such complex ways that the individual is powerless to do anything about them on his own initiative. Only appropriate governmental decisions or services can take the external effects into account and provide the framework within which the decisions of individuals who constitute the community will produce results consistent with the general welfare.

This externality argument serves, in effect, to transform the decision on private vs. public provision of some services into a matter which is primarily technological--the relevant question becomes, who can provide the services in question more effectively in the sense of contributing most to the public welfare per unit of resources employed. And the answer is that where external effects are very important, the public sector is often by far the more efficient supplier.

These three grounds now appear to constitute the rationale of governmental activity. However they must all be subject to a proviso. An extension of governmental activity may yield one or more of the types of benefits which have just been described--but that, by itself, is not enough to justify its introduction. For an addition to a government's role also has its costs. It may involve the use of resources which would otherwise be available to the private sector (a cost which is most important when a national policy maintains a high level of employment and a general shortage of resources). Moreover, if we believe that more government is, at least to some extent, undesirable *per se*, this too must be weighed as a cost. It may smack of the platitudinous to remark that in deciding whether to extend a governmental activity, attention must not be confined either to the benefits or to the costs alone. The extension should only be undertaken if the benefits can, on the basis of explicit analysis, be presumed to be worth the costs. Unfortunately, there are too many illustrations of partisan effusions in which this admonition is ignored, and where either only the benefit or only the cost of a governmental activity is taken into account.

Before leaving this very general discussion of the rationale of public activity it is desirable to consider briefly the determinants of the appropriate spheres of the federal and local governments.

There are two important reasons why some activities can advantageously be left to the federal government. These are closely analogous to arguments which have just been presented.

Regional inequality. Some regions may be considered too impoverished to be left to provide governmental services by themselves. For example, federal aid to education is sometimes advocated on these grounds; they also provided part of the justification for TVA.

Interdependence of different sectors of the country. Here we have the analogue of our last ground for governmental activity--the external effects of the acts of the individual. Similar possibilities hold for the localities which constitute the nation. Because people move about, improved education in City X is, in the long run, apt to redound to the well-being of City Y. Just as street lighting cannot be provided individual by individual, national defense cannot effectively be left to individual communities.

But there are limits to the effectiveness of these arguments. The extension of federal activities at the expense of the local government is not without adverse effects. The federal government may not be as knowledgeable about local conditions as is the man on the spot. Perhaps more important, a uniform national regulation is not easily made sufficiently flexible to adapt well to local requirements. The upshot seems to be that whenever an urban area of decision does not involve serious problems of regional inequity or substantial external effects, powers are best left in the hands of the local government.

EFFECTS OF PRIVATE SECTOR CHANGES ON THE NEED FOR URBAN PUBLIC SERVICES

For some years now we have been witnessing an increase in the supply of urban public services and the scope of urban governmental activity. This development cannot be explained simply as a manifestation of Parkinson's law. Rather, it must be recognized that the need for governmental activity has grown as a result of developments in the private sector, and that changes in the role of the public sector have come about, at least in part, in response to these needs.

In reviewing these developments it must be emphasized that for our purposes the changes which have occurred in the private sector are appropriately treated as exogenous, that is, no attempt will be made to explain their occurrence. Rather, they will merely be described briefly and then used to help to explain developments in the supply of and need for urban services.

Growing Productivity and Wealth

One of the outstanding determinants in the demand for public services (as it is for commodities of any variety) is the magnitude of the flow of real purchasing power coming into the hands of the members of the community. As our income increases so does our effective demand for all sorts of goods and services. In the case of commodities dispensed through the market mechanism, this relationship is completely obvious and scarcely worth mentioning. People just use their higher real incomes to demand more goods and that is all there is to it.

In the case of public services the connection is more subtle and requires some brief discussion. Clearly, people do not normally telephone municipal administrations and offer to pay higher taxes in return for improved services. Yet, as their wealth increases, gaps in the services provided by the government become less tolerable and are increasingly considered grounds for complaint. This is entirely in accord with the standard utility analysis of consumer demand which tells us that the rational individual will seek to keep the ratios of the marginal utilities of the items he consumes proportionate to their cost of acquisition. This means that a substantial increase in his rate of receipt of one type of commodity will normally be accompanied by an increase in his demand for others as well.

As a result, a school-leaving age which was formerly considered acceptable becomes intolerable as the community's wealth increases. Roads and garbage disposal standards are raised. Public pressure for these improvements gradually leads to their effectuation, which in turn may induce the enactment of the appropriate financial measures whereby the necessary resources are collected from the public. Through this process the effects of rising incomes are transmitted to the public sector right along with the increased demands for privately produced commodities.

While this machinery works in the right direction, it unfortunately is even less effective in approximating an optimal allocation of resources than is the market mechanism. To the individual recipient of a public service there is very weak and nebulous connection between the magnitude of the public services received and their direct cost to him. Indeed, there is always the temptation to regard a public service as costless, on the assumption (not always invalid) that it will be paid for by someone else. Perhaps, more often, governmental activities are denounced as too costly because their benefits are given insufficient recognition. There is therefore little reason to expect any close relationship between the magnitude of the pressures for increased public services and the appropriate amounts as determined by the marginal rate of substitution between private and public services when compared with their relative costs. Indeed, there seems to be a very substantial lag in the provision of the additional urban services called for by rising per capita income. For, particularly where local government is concerned, budget restrictions must be considered very real even from the most advanced functional finance point of view. Municipal governments always seem to be operating under conditions of extreme financial stringency and the forces combating increased local taxes or even borrowings are notoriously powerful. As a result, increases in local expenditures are apt to occur only after the absence of some sort of local service, or the low level on which it is provided, has generally come to be considered unacceptable. There is, as a result, some question whether urban expenditure increases have even always offset the erosion of inflation. It certainly seems plausible that while rising per capita incomes have provided strong grounds for an increase in urban outlays, actual urban services have by no means expanded correspondingly. The deterioration of public school systems, constant complaints of inadequate police protection and sanitation services, all seem highly symptomatic.

Rising Urban Population

A second and oft-cited development in the private sector which has made for rising urban expenditures is the secular growth of population. Precisely why increased population should add to the need for public services supplied per capita is not entirely obvious. Offhand it might appear that larger cities offer scope for economies of large scale in the supply of services (both private and public). For example, an information service which transmits its messages by radio should cost substantially the same to operate whether it serves a municipality of half a million or three million inhabitants, though there would doubtless exist the temptation to offer more elaborate services in the latter. Similarly, while libraries would need larger supplies of books to meet the requirements of a more populous area, there is a well-known proposition of inventory theory (the square root theorem) which indicates that for a given level of service the number of volumes need not rise in proportion to the number of users of the library facilities. Simultaneously, it becomes less expensive, per capita, to provide readers with a wide selection of books as the number of users increases.

On the other side of the picture, diseconomies of scale may result from the sheer combinatorial problem which accompanies increases in the number of inhabitants to be served. The standard textbook example of decreasing returns to scale is based on the number of telephone connections which must be provided to service an expanding body of subscribers. To connect two subscribers only one line of connection is needed, whereas to connect three, three possible lines (AB, AC, and BC) are required; six connections are called for by four subscribers, and so on.[1]

---

[1]More generally, if C_n represents the number of connections required by N subscribers, we have the relationship $C_n = C_{n-1} + N - 1$ or $\Delta C_n/\Delta N = N - 1$.

We can see why this is so by noting that the last subscriber (the Nth person) must be connected to each of the N-1 previous users of the telephone service. Professor Vickrey has pointed out that in practice the rather trite telephone case is a poor illustration of increasing cost problems because technological elements yielding substantial economies of scale are also present in the interstage. However, I retain the illustration because it is an easy way to bring out the very real combinatorial problem which is likely to be much more serious in cases involving loose lines of administrative responsibility and their interrelation, rather than a switchboard where the problem is largely technological.

It should be observed that the telephone subscriber problem is not a matter of population density but, rather, an issue arising simply from the sheer number of persons to be served. If we were to envision an isolated city in a desert, each of whose inhabitants lived on a quarter acre of land, then an influx of immigrants could occur without any increase in density within the metropolis--they could settle at its outskirts, each new migrant with his traditional quarter acre. Yet the problem of the telephone company in our city would not be eased by the lack of crowding in on the center city. Indeed, in this case the failure to increase density probably adds to the cost of communication services per capita by requiring even longer lines to be strung as population expands.

It is to be expected that all sorts of analogous problems will arise in a growing city. Co-ordination and control in any expanding organization are, to a large extent, communications problems, and even in a relatively hierarchical system in which not everyone must communicate with everyone else, combinatorial diseconomies of scale are very likely to occur. This is a basic difficulty in all large organizations where centralized control is not easily exercised, and where, as a result, a frequent response is decreased oversight over details on the part of top management and a consequent loss of operating efficiency.

Such diseconomies of scale then are probably widespread in various subtle ways in a large number of the service supply operations of an urban community whose population is growing. A particularly clear-cut case is provided by the provision of transport facilities. The real traffic costs incurred by a rising population are again not a pure matter of population density. If a metropolitan area were to grow by increasing its area proportionately with the number of its inhabitants, the number of roads required to offer the same degree of proximity to each residence would grow in the same proportion. However, assuming a perfectly random movement of traffic, that is, a random set of combinations of sources and destinations, there would be a much greater volume of traffic through the center of the metropolis than at its edges, despite the assumed evenness of population density, and the traffic flow will rise much more than in proportion with the size of population.[2]

Hence, while it may not be necessary to have more roads as the size of the population increases, it will certainly be necessary to have bigger and better roads or suffer the terrible problems of traffic congestion. Indeed the mathematical argument of the preceding footnote suggests why our cities have so far been unable to cope with their traffice problems--for it shows that, to keep the degree of traffic congestion constant, road traffic capacity must rise far more than in proportion with the rate of increase of population, and sheer problems of geography and land availability practically

---

[2]Thus consider an area laid out along a straight line with five population points as shown:

A B C D E

If one automobile travels between any pair of destinations, exactly four cars will pass each of points A and E, seven cars will pass points B and D and eight autos will pass C (AC, AD, AE, BC, BD, BE, CD and CE traffic). More generally, in an N person linear city (where N is assumed to be an odd number), N-1 cars will pass each end point, while $(N-1)^2/4 + N-1$ cars will pass the center point, C because each of the $(N-1)/2$ points to the left of C will have one car going to (or from) each of the $(N-1)/2$ points to the right of C, and, in addition, C will have one car going to (or from) each other point. Hence traffic density at C will rise roughly as the square of N.

preclude such a possibility.[3] Of course, the fact that population tends to cluster and is not spread evenly throughout the city only adds to these congestion problems.

In addition to the combinatorial problems introduced by a growing population which add to the quantity of resources needed to maintain a given level of service to the typical inhabitant, an increase in the size of the urban population has another type of consequence which calls for an increase in public urban service activity. In the previous section it was suggested that one of the basic grounds for the supply of services by the government rather than by private enterprise is the existence of widely diffused external effects which makes it impractical for those involved to co-ordinate their activities to produce a jointly optimal result. As the size of the community increases, these external effects become more and more widely diffused. Social wants--that is, wants of a number of persons which are satisfied by a single act of supply, come to include more and more persons. As this happens, less and less can be left to the public to supply for itself on a voluntary basis. The most clear-cut example of this is the fact that the volunteer fire department's existence is confined to the relatively small town. But more subtle illustrations can be found. Special laws regulating the construction of skyscrapers in order to insure that they do not block off sunshine from the streets is something which is only relevant where a larger population requires larger buildings and where, as a result, the action of the builder has marked and diffused external effects. Similarly, it is only when roads are already crowded that the presence of another motorist makes a substantial difference to others, so that regulations become necessary for their common welfare.

Thus, increasing population adds to the significance and degree of diffusion of the external effects of the actions of all inhabitants of the metropolis and thereby requires increasing intervention by the public sector to assure that social wants are supplied and that externalities do not lead to extremely adverse effects on the community's welfare.[4]

Indeed, the very growth of population itself involves external effects. New residents usually require the provision of additional services and facilities--water, sewage, disposal, road paving, etc., and this is likely to be paid for in part out of the general municipal budget. That is, the additional population may not bear all of the costs involved in its adhesion to the community. As a result, a growing area may attract new residents at an even faster rate than if the social costs of their movement were fully borne by themselves. Sometimes this may be recognized and acted on by local inhabitants who will offer no welcome to "less desirable classes" of migrants. But more often the town booster seems to miss this point and appears to act on the conviction that community growth will reduce the per capita tax burden (by "spreading the overheads") and thus automatically

---

[3]Clearly this statement is only meant to indicate the need, not to describe actual behavior patterns. The Brazer study suggests that per capita expenditure on highways has been inversely correlated with population density. See Harvey E. Brazer, City Expenditures in the United States, occasional paper 66, National Bureau of Economic Research, New York, 1959, pp. 25-29.

[4]*Ibid.*, shows that there is no significant growth in per capita urban expenditure as population expands, and that where there is a significant relationship it is negative--urban expenditures lag behind population growth. This interesting fact does not conflict with my argument that population may involve substantial external effects by placing a disproportionate burden on old residents. Part of this burden, clearly, may be a deterioration in the quality of municipal services.

yield universal net benefits. In such cases governmental intervention may perhaps have proceeded in precisely the wrong direction.

Technological Change

A third change in the private sector which has induced increases in public activity has been the nature of recent technological change which has also led to marked increases in external effects. Many of the standard illustrations of the external economies argument are to be found here--the dangers of air pollution arising from automotive transport and from modern factory processes, the problems of traffic control arising from the invention of the automobile, the noise and danger introduced by the location of airports in close proximity to the metropolis (and the inconvenience resulting when it is located too far away) are all, at least partly, examples of external effects underlying important municipal problems, which have resulted from the technological developments of this century and which our growing technological knowledge continues to produce.

Not all technological change must necessarily increase the magnitude and significance of the external effects of people's activities, and hence require expanded governmental operations. But it it highly plausible that innovation is heavily biased in this direction. To see why this is so we must recognize that the absence of external effects is assured only when the welfare of each individual is independent of the activities of others. In such a world each man is, indeed, an island. External effects are a consequence of the interdependence of the welfare of the members of the community. Now it it plausible that some inventions--say the development of locks--served to reduce external effects, and hence the need for urban services--in this case, the demand for police protection. More recently, the development of home appliances such as washing machines and driers, or home workshops, have tended to permit greater isolation to the individual (at least until a repairman is needed).

But most innovations tend to have at least one of the following three characteristics:

1. They can serve many people simultaneously--an extreme case being any weather information, communications transmission and other benefits obtained from an artificial satellite program.
2. They are sufficiently complex to require the co-ordinated contributions of several people or organizations, and it is difficult to impute completely and unambiguously the value of the final product among those who contribute to its realization.
3. They have unintended by-products. They create noise, a variety of wastes, pollution, crowding of roads, etc.

It is clear that any innovation having one or more of these characteristics is likely on balance to add to the prevalence of external effects. Hence my surmise on this point, that innovation is inherently biased toward the production of externalities and that it therefore has and is likely to continue to add significantly to the need for public services.

A particularly acute problem of this variety is of some special interest--the difficulties of the public transport systems. In most metropolitan areas public transport has come under increasing financial pressure of such magnitude that in some areas it is threatened with extinction. The automobile has enabled more and more people to supply their own transportation. Public carriers have responded to the resulting loss in passenger traffic in the most obvious and ultimately, perhaps, the most disastrous manner. They have raised fares and reduced service in an effort to make ends meet. The inevitable result is a still further reduction in the number of users which, in turn, only calls forth ever higher rates and poorer service and so on, apparently ad infinitum. This sort of cumulative process is at the heart of some of the most critical of current urban problems and it will therefore be examined in somewhat greater detail presently. For the moment we need merely note that public intervention is very likely to be required if the process of deterioration is to be arrested. The reason once again lies within the family of external effects cases. As in a Greek drama, each character is (at least by himself) powerless to halt the course of the tragedy, and his best efforts to cope with the situation only increases the difficulties which beset the others and himself. Thus, everyone--the reluctant passenger and the unhappy transport official alike--is forced into decisions whose adverse effects fall on others and are then ultimately transmitted back to themselves. Perhaps we can generalize by saying that disequilibrium in any social or economic situation is very likely to involve a complex of externalities. This point will be expanded upon presently.

Movement to the Suburbs and Urban Blight

Another major exogenous change which has influenced the need for public participation in the urban economy is the growing popularity of suburbia and exurbia. To some extent this has served to offset the rise in population which was previously discussed. But perhaps more important than the effect on the total number of inhabitants have been the economic effects of the resulting migration. For it would appear that the new class of suburbanites is drawn largely from the middle (and perhaps upper) income families, whether because the taste for country living is partly income determined or because it is, relatively, expensive to satisfy this preference.

At any rate, the consequence seems to have been a reduction of the average income level of those who remain in the metropolis, and from this a number of important consequences have followed. There has been some direct physical deterioration as formerly prosperous areas have been taken over by poorer residents who are less able, and perhaps less interested than former inhabitants in maintenance efforts and outlays. In addition, some business firms have found it expedient to follow the population movements. A few offices and organizations whose employees are primarily professional--research organizations are noteworthy in this respect--have apparently concluded that it is easier to recruit and retain staff when attractive residential possibilities can be found in close proximity to the location of new operations. More obvious has been the effect on retailers. Expansion has become increasingly difficult for downtown department stores, and in some cases, they have been forced to close down operations altogether. Instead, suburban shopping centers have followed the public's purchasing power out of the heart of the city. The phenomenal growth of this institution hardly requires elaboration or documentation.

The exodus of the middle classes and of some classes of business firms has created financial problems for the municipality. Given the structure of taxes and the level of tax rates, declining income levels, sales rates, property values and business values mean a restriction of the funds available to the government. At the same time, these changes make it more difficult to increase tax rates. Moreover, the relative increase in poverty is likely to require increased urban expenditures for welfare services, unemployment relief, and other related purposes. In sum, the exodus of income to the suburbs imposes on municipal governments a double pressure--it makes it more difficult for them to obtain resources and it tends, simultaneously, to force them into heavier expenditures.

Here we have a second important illustration of a cumulative process, a movement whose difficulties feed upon themselves. Migration of the middle classes helps to produce urban blight, to reduce the availability of shopping facilities and other amenities and it is likely to force the government to reduce the quality of its services--it may lead to deterioration of the urban public school system, of the public hospital services, of the effectiveness of the police force, the sanitation services, etc. But all of these unfortunate consequences further reduce the attractiveness of urban living and may induce still more of those who can afford it to join the exodus. It is not uncommon, for example, to find confirmed city dwellers leaving because they cannot arrange for satisfactory schooling of their children within the city. The rest of the story should now be obvious. Increased migration accelerates blight, and increasing blight hastens migration. Successive rounds of this procedure constitute what may well be today's major problem for the public sector in the cities, for surely the deterioration of our urban centers is something no one can condone or desire. And by now it should be clear that it is a problem for the urban sector because it is another external effects case, a difficulty from which the individuals involved cannot extract themselves by their own unco-ordinated efforts.

This largely completes our examination of the effects of recent exogenous changes in the private sector on the need for urban expenditures and regulatory activities. The nature of some of these government acts still remains to be discussed. But first, since it plays so important a role in the considerations which have just been listed, it may be appropriate to digress and attempt a formal analysis of the process of cumulative deterioration which characterizes both the decline of public transport and the economic problems of the center city.

Digression: The Theory of Cumulative Deterioration[5]

Cumulative processes readily suggest the simplest and most standard of dynamic sequences--those which can be described by first order difference or differential equations. If we can visualize some index of urban blight and deterioration at date t which we symbolize by B_t, what is asserted in our model is that

$$B_t = g(Y_t), \quad dg/dY < 0. \tag{1}$$

That is, the state of blight at time t is a decreasing function of the level of per capita income, Y_t at that date. In turn, the exodus of more well to do inhabitants is affected (adversely) by the state of blight. Here we may profitably consider two alternative hypotheses: first that blight adversely affects the level of per capita income, or, alternatively that it increases the rate of fall of per capita income. In the former case we obtain

[5]The reader who is neither enlightened nor edified by a mathematical restatement of an economic problem is well advised to omit this section.

$$Y_{t+1} = G^*(B_t)\,, \qquad dG^*/dB < 0 \tag{2*}$$

and in the latter case we have

$$Y_{t+1} - Y_t = G^{**}(B_t)\,, \qquad dG^{**}/dB < 0\,. \tag{2**}$$

Combining equations (1) and (2*) we obtain

$$Y_{t+1} = G^*[g(Y_t)] = F^*(Y_t)\,, \qquad dF^*/dy > 0\,. \tag{3*}$$

Alternatively, combining equations (1) and (2**) we get

$$Y_{t+1} - Y_t = G^{**}[g(Y_t)] = f(Y_t)\,, \qquad df/dy > 0\,,$$

that is,

$$Y_{t+1} = f(Y_t) - Y_t = F^{**}(Y_t)\,. \tag{3**}$$

Generalizing (3*) and (3**) we have,

$$Y_{t+1} = F(Y_t)\,. \tag{3}$$

This is a standard first order nonlinear difference equation. We will deal first with the general case (3) and later return to (3*) and (3**). Two possible forms of relationship (3) are represented graphically by means of a phase diagram. Figure 1 represents the unstable case which is most closely in accord with our intuition. If per capita income is ever driven below some critical level Y_K, say to point Y_o, the march toward doom becomes irreversible in the absence of outside intervention. Because the point on the graph of $Y_{t+1} = F(Y_t)$ corresponding to Y_o (i.e., point A) lies below the 45° line (where $Y_1 = Y_o$), we find that $Y_1 < Y_o$. That is, blight increases, and per capita income falls. In the next period, starting from the lower income level Y_1, per capita income is driven down to the still lower level, Y_2, and so on. Income falls along the path A B C D ... and there is nothing to prevent it from declining without limit.

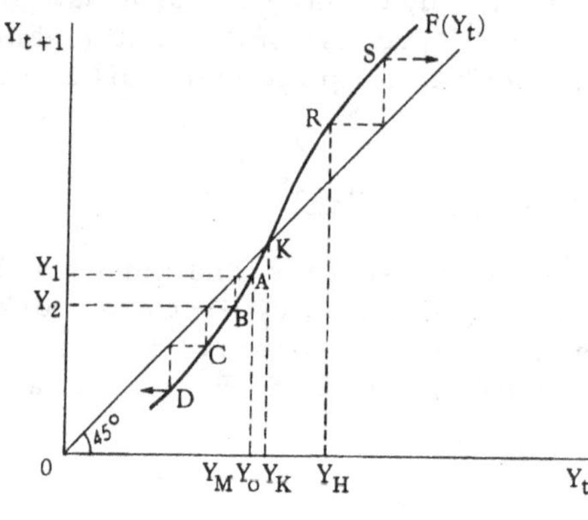

Figure 1.

By contrast, Figure 2 represents a stable (though not necessarily a more happy) situation. Here, from Y_o per capita income also declines steadily, but there is a limit to the amount of the decline. Y_t asymptotically approaches the limit level Y_k and never falls further than this. This latter, stable case may at first glance appear more desirable from the social point of view. For at least it offers some degree of reassurance to the declining community that the situation cannot deteriorate beyond any preassigned limit. But this advantage is apt to be illusory.

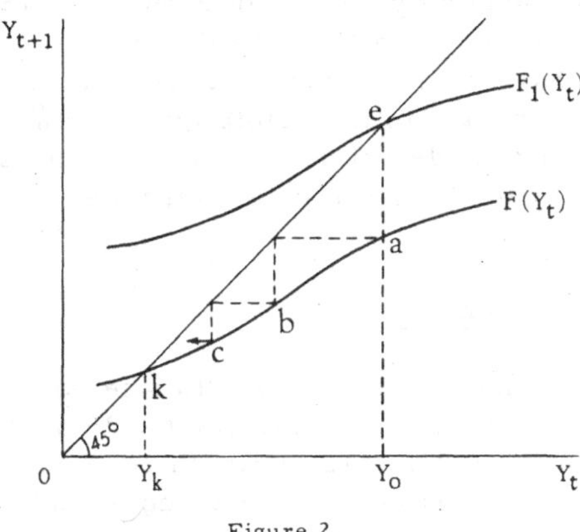

Figure 2.

First of all, the limit level of per capita income, Y_k may be very low indeed, and offer small grounds for satisfaction. Moreover, it is inconceivable that in the unstable case (Figure 1) Y_t will ever be permitted to fall to zero. Long before the economy of the area threatens to collapse altogether something radical is virtually certain to occur whether by design or by force of circumstances, though it must be admitted that the advent of chaos must also be considered as a possibility in such a situation. In sum, then, it is economy 2 which is perhaps in danger of stagnating in a chronic nadir of lethargy, but this luxury will be impossible in the unstable situation.

The latter has yet another important advantage over the stable case. Suppose, in Figure 1, per capita income has fallen, say to level Y_M, but by outside assistance or heroic internal efforts it is somehow temporarily increased, say to Y_H--above the critical level Y_K. According to the diagram the situation would then suddenly change completely. From that time on it could safely be left to take care of itself, income would improve automatically by moving upward and to the right along path R S . . . . Thus, a sort of pump priming may be a very real possibility for the policy maker in the unstable case. But no such hope exists in a stable situation. In Figure 2, no matter how high per capita income is raised, in the absence of a major structural change in the system, it will lead down again toward Y_k.

We may also ask what determines whether the situation will be stable or unstable. This depends on whether at point K (or k) the slope of the graph of $Y_{t+1} = F(Y_t)$ is greater or less than that of the 45° line. That is, the situation will be unstable if at that point $dF/dY_t > 1$ and stable if $dF/dY_t < 1$. Returning now to our two special hypotheses represented by (3*) and (3**) above, we note that in the former case $dF^*/dY_t = > 0$. That is, the model in which blight adversely affects the level of per capita income is necessarily unstable.

On the other hand, in the other model summarized by (3**), we have

$$dF^{**}/dY_t = df/dY_t - 1$$

whose sign is ambiguous. If a fall in income produces a relatively negligible increase in blight and this, in turn, has only a minor effect on the rate of fall of per capita income (both dg/dY and dG^{**}/dB small), it follows that $dF^{**}/dy = (dG^{**}/dB)\,(dg/dy)$ will also be small. Then and only then will the situation be stable.

In practice it may be suspected that both the difficulties of public transportation systems and the exodus to suburbia represent stable cumulative processes and there may be some evidence that both of these are starting to level off in some places, often toward very low and unsatisfactory levels. What probably happened was that an initially satisfactory and stable equilibrium level[6] such as that represented by point e in Figure 2 was moved as the result of exogenously produced shift in the F curve from, say $F_1(Y_t)$ to $F(Y_t)$. Such a shift would have been produced by the advent of widespread ownership of private motor cars in the public transportation problem case, and in the case of the urban deterioration problem the shift would have resulted from an autonomous change in public attitudes toward suburban living.

The Nature of Appropriate Public Intervention

It would clearly be inappropriate and, indeed, foolish to offer detailed recommendations for public actions on the basis of our relatively abstract and general discussion of changes in the public sector which require attention by the government. However, some pertinent tentative comments can be offered.

First of all, it seems clear that almost all of the trends which we have examined call for increases in public activities and outlays, very likely on a substantial scale. Increasing population density, pollution problems, the crises in transportation and the overall urban economic structure clearly require increased public services, more stringent regulations, and greater intervention generally. A municipal government which is committed to a hands-off policy can only expect these problems to grow increasingly critical of their own accord. The necessity for such an extension of governmental operations must be deemed unfortunate by those who, like myself, consider it inherently undesirable to expand the scope of operations of the public sector. But the costs of government intervention must be weighed against the costs of its absence, and I think in this case the answer is clear--the rising scale of urban public activities in recent years is not only justified, it is probably vastly insufficient.

A second point requires emphasis in this connection. The presence of external effects and other grounds for increased governmental intervention need not constitute a license for petty bureaucrats and others to impose their views of virtue and good living on a recalcitrant public. It is perfectly consistent to believe in public sovereignty and to advocate an increase in governmental activities in particular circumstances. But in such cases it is appropriate for the government to intervene on public sovereignty grounds precisely because the situation is in some way frustrating the desires of the public. Government intervention must then be carefully tailored in a manner designed to restore public sovereignty, not to frustrate it further. If, for example, it is decided to subsidize public transport, this should be done only if it is estimated by whatever crude means are possible, that this will yield a greater advantage to the public (in terms of the preferences of the inhabitants themselves) than will any alternative use of the resources involved. The fact that some public official feels that public transport is a good thing in terms of the American way of life and true aesthetic virtue is entirely beside the point. This is, of course, a value judgment. But it is important to realize that there is a middle ground between the extreme interventionist who feels he knows what is good for

---

[6]This is very likely to have been a moving equilibrium, with expansion imparted to it largely by continuing exogenous changes which are ignored in this section because they do not affect the issue.

others better than they do themselves, and the diehard who would keep the government down at almost any cost.

A final comment on the nature of appropriate increases in governmental urban activities stems from our discussion of the cumulative deterioration process. It should be clear that these problems are serious and that their solution is likely to require extremely radical measures. Cities which are turning into vast slums and are rapidly choking themselves on snarled traffic cannot expect that an additional freeway or a few low-income housing projects will make a significant change in the problem. We have seen in the preceding section that in a stable cumulative process such acts are likely to be palliatives which at best temporarily arrest the course of deterioration. The fact that freeways seem frequently to have turned out to be obsolete before they were completed and that some of the postwar housing projects are already acquiring slum characteristics seems to confirm this view. Our analysis has indicated that only a really radical change which modifies the very structure of the situation may offer real hope. For example, it may be necessary to undertake acts as extreme as the complete banning of privately owned passenger cars from downtown streets to cope with the traffic problem. However, I must reluctantly take a pessimistic view of the likelihood of such courageous and effective programs. Politicians seem to vie with one another in promising to continue business as usual and to shun all activities which are unhallowed through the respectability imparted by ancient usage. We may well anticipate the reward of such timidity in situations which require boldness and imagination.

ALTERNATIVE MEANS OF IMPLEMENTING PUBLIC SERVICE POLICY

Once it has been decided by a public body that some additional urban public service is appropriate, it remains to determine the most effective means for supplying it. For example, suppose it is decided that sidewalks should be built in a newly developed area. The government may undertake to build them itself, using municipal employees and facilities; or it may decide to hire a private firm to undertake the construction. Still another alternative is for the government to offer subsidies to developers who install sidewalks during the course of their building activities. A final possibility is that the government may simply *require* the developers to provide them. All four approaches are clearly potentially capable of achieving the desired results and each of these methods is employed in different circumstances. Schools are typically operated and, to some extent, maintained directly by the municipal governments. Streets are frequently built and repaired by private contractors paid out of public funds. In some cases, at least, urban redevelopment activities by private firms are strongly encouraged by grant of special privileges or more direct subsidies and, finally, houseowners are required to remove snow from their sidewalks and provide a variety of other services to the public in general.

The principles on which selections are made from among these alternatives are by no means clear. One suspects that time-honored practices, fortuitous events, political considerations and even elements of cupidity have played significant roles in the patterns which are currently prevalent. Perhaps more interesting for present purposes than a descriptive and explanatory examination of present decision methods is a normative discussion of the principles on which such decisions ought to be based.

Unfortunately, it would appear that relatively little can be said on this subject in general terms. Appropriate decisions in these matters can only be made case by case, in terms of the particular circumstances. This will become clearer as several general principles are enunciated because it will be seen that the light which they cast is very feeble indeed.

Efficiency. There is a prima facie case for employing the means which can get the job done with the smallest use of resources. For example, if a developer may be expected to have available at his building site cement mixers and other necessary equipment this may make it inadvisable for anyone else to incur the effort of organization required to put in sidewalks. On the other hand, if considerable economies of large scale can be achieved by using a single contractor to build all sidewalks the task should be taken out of the hands of individual developers. Snow removal may naturally fall to the municipality itself if it has an organization which normally cleans the streets, because both street cleaning personnel and equipment are readily transferred to snow removal work and would otherwise remain idle during a period when streets are covered with snow.

Repercussions on the private sector. It is also relevant to take into account the secondary effects of these decisions on the private sector. For example, there is much to be said, other things being equal, for having work done by private firms either by government contract or under partial or indirect subsidy. For such an increase in private activity is likely to provide more stimulation to business buoyancy and willingness to invest than is an equivalent activity undertaken directly by the government. It may also be helpful in attracting firms to the area if that is desired (though, as pointed out earlier, the advantages of town boosting of this sort are not always as clear-cut as they appear at first view). Of course, in inflationary periods, stimulation of business may not be pressing or even desirable, but, by and large, counterinflationary action is a matter more appropriate for federal than for local concern.

The same argument urges caution in deciding to have the government take over an activity which was formerly in private hands. Such a course is sometimes desirable, on balance, but it is apt to be discouraging to private enterprise unless private enterprise had already intended to withdraw from the area.

This is not to deny that publicly inaugurated urban services can stimulate business activity. They can and do so in many important ways. The provision of roads, and other facilities are clearly extremely important, and, indeed, private enterprise cannot be relied on to provide them adequately because of the external benefits which these services offer and for which their supplier cannot charge on any practical basis. The view that the municipality should see to it that such services are supplied is not in question here. Rather, what is asserted is that the effects on the private sector may make it appropriate to have such items supplied through some private agency which is hired or subsidized by the public authorities.

In general, then, it would appear from this argument that other things being equal it is per se preferable to have services supplied through private agencies than by the municipalities directly. Of course, other things are almost never equal and so, considerations of relative costs, and other elements which remain to be discussed, make it desirable that the supply of a considerable variety of services remain entirely in governmental hands.

Psychic effects. In some cases public or private operation may be considered a virtue in and of itself and may be deemed desirable because it offers satisfaction to the public which is essentially no different from the economist's point of view than is the building of a large monument or the maintenance of a guard of honor for ceremonial occasions. People may take special pride in a municipal operation feeling that it belongs to them rather than to "the interests." It is surely on these grounds that nationalization has played so important a role for the British labor party.

In urban activities this consideration is not as remote as it may seem. It is by no means obvious to me that continued municipal operation of the public schools offers a clear cut advantage in efficiency and maintenance of high standards as against the potential performance of a privately operated school system operating under government license and control, and financed by the municipalities.[7] Though the analogy is far from precise, it is suggestive that the performance of the private colleges and universities is not significantly less (or more) satisfactory, on the whole, than that of public institutions. The strongest arguments for the maintenance of the present form of the public school systems are the fact that it is already operating (and change is always costly), and, what is more relevant at this point in our discussion, that there is a strong and widespread sentimental attachment to its current mode of operation. Even where school systems have deteriorated markedly, this is a serious and significant ground for hesitation before undertaking any radical changes, because the social cost of the destruction of the symbol of public operation may be a very real cost indeed.

Equity considerations. An important element in the decision as to means of supplying a public service is the question of distributive justice. Whether developers should be subsidized to induce them to supply sidewalks, or forced to supply them at their own expense (or that of the purchasers of the properties they have constructed) is very clearly such an issue. Having raised this problem I shall say no more about it. Economists have been notoriously unsuccessful in their attempts to supply helpful suggestions on matters of distributive justice and this paper is no exception.

Weighing the alternatives. Probably, the foregoing list of considerations is seriously incomplete but there is little purpose to proliferation of its entries. There is problem enough in assigning weights to those which have already been mentioned, or even in describing effective methods for obtaining the measurements which are involved. It is easy to say that if a public service can be supplied most cheaply by a private contractor and if this method of operation involves no blow to civic pride, he should be hired to provide the service. But what of a case where the municipality can operate an activity at a considerably lower real cost but where added government activity is likely to affect business operations adversely?

---

[7] Let it be clearly understood that this is not an argument for government subsidy of church affiliated private schools whose operations fall outside the range of direct government supervision.

How should these effects be measured and compared with one another? It is, unhappily, easier to raise these questions than to answer them.[8]

CONCLUDING REMARKS

Questions relating to decisions on the public supply of urban services have been examined at some length and an attempt to develop some theoretical basis for these decisions has been undertaken. Much of what has resulted should not be dignified by being referred to as a theoretical structure and some of it is sheer taxonomy. It seems plausible that this is in part an inescapable result of the very broadness of the problem. In the course of this paper the source of the need for public services has been recapitulated. The discussion went on to apply this analysis in an examination of the effects of recent changes in the private sector on the desirable range of activities of the public sector. From this there emerged the view that in some areas a very considerable expansion in the range of decision of the municipalities is urgently required--an expansion greatly exceeding that which has already occurred. Perhaps further and more directly empirical study of these matters can convince those involved that they must act effectively before the problems become even more acute.

---

[8]One surmise can be offered here. It seems plausible to me that an optimal program, however determined, will agree with present practices in at least this respect: that each of the methods of affecting public service policies will be employed to some extent. That is, it will not be desirable to rely exclusively on direct municipal operation, or entirely on subsidies to private firms, or on any other subset of our four possibilities. Essentially, this view is based on the feeling that there are diminishing returns to excessive reliance on any one of these means. For example, direct municipal operations are only likely to yield serious disincentive effects for business if they are relatively large, and this consideration is probably of little relevance so long as direct public activity is fairly limited. Examination of each other mode of operation suggests that its disadvantages, inefficiencies and irritant effects are also apt to grow disproportionately with its scale of operations. The desirability of a "balanced program" follows directly from this argument.

2

Changing Patterns of Local Urban Expenditure

by Allen D. Manvel

"Public expenditure in urban communities" might be broadly interpreted to involve spending by states and the federal government as well as by local governments. Every hamlet in the United States includes a post office, and every sizable city has some state government institutions or activities and some federal government offices, in addition to being served by a municipal government and—usually—a county and other separate local governments.

I shall apply no such comprehensive definition. Instead, the following discussion will deal only with expenditure made by local governments. This specific and somewhat measurable focus permits the observation of developments and prospects that might be obscured in a broader context.

Such an approach makes the task of quantification somewhat less unmanageable, though inter-area variations still get swallowed in nationwide aggregates and averages. This approach also has the virtue of being concerned with governmental services provided for particular local areas as such, rather than embracing also the range of benefits, services, and facilities which are available on a nationwide basis (e.g., national defense) or to the citizens of entire states (e.g., public universities, mental hospitals, etc.).

The dividing line is not absolute or uniform. Under our federal system there is considerable variety in the state-local pattern of responsibility for providing certain services, such as public welfare, health, and highways. There is even greater variation in the form and extent of state participation in the financing of various functions commonly performed by local governments.[1] The latter kind of diversity does not directly interfere with our analysis, however, for we shall deal with local government expenditure in terms of gross amounts, however financed. The sums involved are thus

---

[1] For further detail on this point, see Selma Mushkin's essay, "Intergovernmental Aspects of Local Expenditure Decisions," pp. 37 to 61 below.

far larger than revenue from local taxes. Nationwide, in 1960, locally imposed taxes yielded $18.1 billion, but gross local government expenditure was more than twice this great, having been financed to a considerable degree from local non-tax revenue, state and federal grants, and borrowing, as well as local taxes.

The focus on local government expenditure, it may be conceded, involves an element of artificiality. Urban areas are tremendously affected by federal and state spending, and a case might be made that state and local amounts are so closely interrelated that aggregates or averages pertaining only to local government have but limited significance. I believe, though, that this would be an overstatement. With all the exceptions and variations to be found, there is still a broadly prevailing pattern of locally assigned governmental responsibilities. This is particularly true for so-called "urban" services such as fire protection, water supply, and sewerage, but it also applies to the provision of public elementary and high schools and in considerable degree to numerous other functions. In any event, local governments are on the front line of exposure to most of the needs of urban communities for domestic public services, and are significant instrumentalities for meeting many of those needs. Even for domestic functions handled mainly by states or the federal government, future prospects are likely to be much affected by developments in local government operations and finances.

One might explore various kinds of "patterns" of public expenditure in urban communities--for example, from the standpoint of the types of governments involved (cities, counties, school districts, special districts, and, perhaps broadly, states and the federal government); the forms of expenditure (personal services and other current operating costs, construction, land acquisition, public assistance); the kinds of financing involved (from user charges, benefit taxes, borrowing, intergovernmental revenue, and local general taxes); or, perhaps, from the standpoint of geographic incidence, considering the relation of central to outlying portions of metropolitan areas and the place of smaller separate urban communities in the total picture. I shall not deal with any of these kinds of patterns but, instead, shall discuss local government expenditure in urban communities mainly in terms of the various functions being served.

COMPOSITION OF LOCAL URBAN EXPENDITURES

Local government direct expenditure in 1960, according to Census Bureau data, totaled $38.7 billion, or slightly over $200 per capita.[2] Table 1 shows the functional distribution of this total.

For a number of reasons, however, these over-all figures do not closely reflect the prevailing pattern of local government expenditure in urban communities. To develop a closer approximation of that pattern, on a per capita basis, the following adjustments may be applied to the basic Census data: (1) Eliminate amounts for those components which in

---

[2]U.S. Bureau of the Census, Governmental Finances in 1960. Except where otherwise indicated, local government aggregates in subsequent discussion and tables are from this source and similar earlier annual Census reports, except that 1957 data are from the 1957 Census of Governments. Except for 1957, these local data are survey-based estimates subject to sampling variation. The Census figures pertain to local fiscal years ended in the calendar year cited.

Table 1. Local government expenditure, by function, 1960

Function	Amount ($ million)	Per cent	Per capita ($)
Total	38,677	100.0	215.68
All general expenditure	33,931	87.7	189.22
Education	15,162	39.2	84.55
Institutions of higher education	346	0.9	1.92
Local schools	14,816	38.3	82.62
Streets and highways	3,358	8.7	18.73
Public welfare	2,183	5.6	12.17
Public assistance	1,503	3.9	8.38
Other public welfare	680	1.8	3.79
Health and hospitals	1,899	4.9	10.59
Sewerage	1,103	2.8	5.97
Sanitation other than sewerage	624	1.7	3.66
Police protection	1,612	4.2	8.99
Fire protection	995	2.6	5.55
Housing and urban renewal	850	2.2	4.74
Local parks and recreation	770	2.0	4.29
Airports and terminals	473	1.2	2.64
Natural resources	347	0.9	1.93
Libraries	261	0.7	1.46
Correction	253	0.7	1.41
Interest on general debt	1,134	2.9	6.32
General administration	1,459	3.8	8.14
General public buildings	413	1.1	2.30
All other	1,034	2.7	5.77
All utility expenditure	4,066	10.5	22.67
Water supply systems	1,881	4.9	10.49
Electric power systems	1,244	3.2	6.94
Transit systems	750	1.9	4.18
Gas supply systems	191	0.5	1.07
Liquor stores expenditure	115	0.3	0.64
Employee retirement expenditure	565	1.5	3.15

most communities are performed privately rather than by local governments--i.e., liquor stores, and utilities other than water supply. (2) Eliminate the amount for natural resources expenditure, which applies mainly to rural areas. (3) Eliminate amounts for those subfunctions which widely or predominantly are handled directly by state governments, rather than through local governments--i.e., public assistance and institutions of higher education. (4) In lieu of including "employee retirement expenditure" (which consists of payments of benefits and withdrawals by locally administered retirement systems), take account of local governments' employer contributions to local and state-administered systems (which in the basic Census framework are treated as intra- and inter-governmental transactions); allocate such contributions in proportion to local government annual payrolls for various functions. (5) Also allocate functionally the bulk of local expenditure for interest on general debt, according to the purpose distribution of outstanding long-term local debt as reported by the 1957 Census of Governments. (6) For those local government functions which are substantially limited to urban areas (sewers, fire protection, housing and urban renewal, local parks and recreation, refuse collection and street cleaning, libraries, and nonhighway transportation), calculate per capita amounts by reference to "urban" population only, as distinct from the nationwide total of population.[3] (7) Take account also of the strong influence of urbanization upon expenditure for police protection, by recalculating on a basis which presumes twice as much local spending per capita for this function in urban areas as in rural areas.

As this estimating methodology illustrates, some local government facilities and services are substantially limited to areas of concentrated population, while others--including public schools, roads, and public health and hospital services--tend to be available more uniformly insofar as population concentration is concerned. The local police function seems to be intermediate, involving considerably more intensive provision in urban than in rural areas.

When the calculations described above are carried out, "basic" local government expenditure of urban areas in 1960 is found to have averaged about $212 per capita. If it is presumed that, with 70 per cent of the nation's population, urban areas directly accounted in 1960 for three-fourths of the gross national product, the per capita product amount for such areas is found to equal $3,013. In relation to that sum, the $212 per capita of basic expenditure by urban local governments amounted to approximately 7 per cent. If urban areas are credited with 70 per cent of the nation's gross national product in 1952 (when they had about 65 per cent of the population), the per capita urban product is estimated at about $2,370. Basic urban public expenditure in that year--derived in the same manner as for 1960--is estimated at about $126 per capita, or 5.3 per cent of the product amount.

These estimates indicate a considerably faster rise in urban governmental expenditure--up about 6.7 per cent a year on a per capita basis--than in urban product, which on a per capita basis showed an average rise of around 2.7 per cent annually between 1952 and 1960. Table 2 gives a distribution by function of per capita expenditure amounts calculated in the manner which has been described, together with a summary measure of approximate annual rates of change.

The discontinuity of historical data sources--including changes in definition of "urban" population as well as the limited detail of earlier statistics on local government finances--makes it impossible to construct directly corresponding estimates of urban local expenditure for remote historical periods.

The most notable development which appears from the 1952-60 comparison, of course, is the considerably faster rise in expenditure for public schools than for other local government functions. This can largely be attributed to the dramatic expansion which has been taking place in school age population--a phenomenon which is more fully considered in subsequent discussion of local expenditure for education.

A considerable portion of the changes between 1952 and 1960 can, of course, be traced to inflation. The implicit price deflator for purchases by state and local governments rose about one-fourth between 1952 and 1960--i.e., at an average annual rate approaching 3 per cent. General indexes of wholesale and retail prices went up considerably less than this, but average pay rates for full-time local government employees (with due

---

[3] The figures involved, from the Census of Population, are: total 1960 population, 179,323,000; 1960 urban population, 125,269,000; total 1952 population, 156,427,000; 1952 urban population (estimated), 101,948,000.

Table 2. Basic local government expenditure of urban areas, by function

Function	1960 estimates: Per capita amount	1960 estimates: Per cent	1960 estimates: Per $1,000 of urban product*	Average percentage annual increase in per capita amount, 1952-60
Total	$211.97	100.0	$70.34	6.7
Public schools	87.88	41.5	29.17	8.6
Essentially urban functions, total	59.13	27.9	19.62	5.2
Water supply	15.21	7.2	5.05	5.8
Sewerage	10.02	4.7	3.33	5.9
Fire protection	8.26	3.9	2.74	4.1
Housing and redevelopment	7.80	3.7	2.59	1.8
Local parks and recreation	6.31	3.0	2.09	8.6
Sanitation other than sewerage	5.13	2.4	1.70	3.0
Airports and terminals	4.28	2.0	1.42	9.4
Local libraries	2.12	1.0	0.70	6.2
Other local government functions, total	64.96	30.6	21.55	6.0
Streets and roads	19.81	9.3	6.57	4.5
Health and hospitals	11.12	5.2	3.69	5.6
Police protection	11.05	5.2	3.67	6.3
Public welfare	3.87	1.8	1.28	9.7
Correction	1.44	0.7	0.48	6.3
General administration	8.46	4.0	2.81	5.6
General public buildings	2.30	1.1	0.76	5.4
Miscellaneous and unallocable	6.91	3.3	2.29	10.4

*Crediting urban areas with 75 per cent of all GNP.

allowance for an increase in the proportion of the total representing school system personnel) moved up by about 40 per cent during the eight-year interval, or at an average annual pace of 4.3 per cent.

Census Bureau data on public employment, though somewhat less detailed functionally than the financial statistics, tell a similar story for the period 1952 to 1960.[4] They show a considerably more rapid rise in public school employment than in local government employment for other functions, reflecting the faster increase in school-age population and public school enrollment than in total population. For such functions as local streets and highways, water supply and fire protection, the upward trend in local employment merely paralleled population growth, but for other functions it was somewhat more strongly upward.

PROSPECTS FOR THE FUTURE

Some of the factors that will influence the prospective course of urban local expenditure are self-evident. As in the past, growth of population, urbanization around major cities, economic growth, economic fluctuations, and price level changes will all play an important role. One may also reasonably expect a continued tendency for the pay rates of local government employees to increase more rapidly than the general price level, on the ground that many of these employees are in the relatively skilled occupational groups which are in short supply.

[4]See U.S. Bureau of the Census, Census of Governments, 1957, and annual reports, State Distribution of Public Employment.

But it is necessary to delve more deeply to consider the likelihood of important shifts in the relative scale of various functional components of local government expenditure. Historical U.S. totals for major categories suggest that there has been a rather high degree of stability in the expenditure ranking of various local government functions; they have largely moved in a parallel fashion. The more obvious departures from prevailing trend are traceable to institutional and financial developments--most notably of all, perhaps, the shift of primary responsibility for public welfare from local governments to the states and federal government, which began in the 1930's. Looking ahead, then, it is essential not merely to note such matters as the prospective trend in population, but also to consider the prospects for institutional changes that may alter the role of local governments with regard to particular functions. With this in mind, let us review the several major fields of local public expenditure, and consider factors that may influence their scale in the years ahead.

Education

It has been estimated above that spending for public school purposes made up about 41 per cent of all "basic" local government expenditure in urban areas in 1960--far more than any other function. We have also noted that this category has been increasing considerably more rapidly than local expenditure for other purposes. What are the future prospects?

A somewhat slower rate of increase is suggested by population trends. Nationwide, during the past decade, the population aged 5 through 17 rose at an average annual rate of 3.6 per cent; for the decade of the 1960's, the average rate is expected to be only about half this great. Looking solely at urban population, one finds for the decade just ended that the 5 through 17 group increased at an average annual rate of 5.4 per cent, while other urban population was growing only 1.9 per cent annually.[5] This sharp contrast in growth rates is diminishing. For the rest of the present decade, the urban school age group apparently will increase at about the same pace as urban population as a whole.[6]

For a number of reasons, however, this development will not necessarily bring the trend in urban school spending into line with trends for other local government functions. In the first place, most of the easing of anticipated growth rate will apply at the elementary school level. Projections of public high school enrollment show the same strong upward trend for this decade as took place in the 1950's.[7] Accordingly, there will be a growing proportion of pupils in the high school grades, which typically have smaller classes and higher teacher salaries than elementary grades, with resultant higher costs per pupil.

Another factor, more difficult to evaluate, relates to private and parochial schools. During the past decade, their proportion of all pupils enrolled in kindergarten through the 12th grade rose from less than 12 to more than 15 per cent. (No doubt the percentage is materially higher for urban areas.) The Office of Education has estimated that the nationwide

---

[5]Computed from U.S. Bureau of the Census, 1960 Census of Population, General Population Characteristics, U.S. Summary, pp. 1-148.

[6]U.S. Department of Health, Education, and Welfare, Health, Education, and Welfare Indicators, 1961 Edition, p. 35. The comment is based on midpoints among four alternative projections.

[7]*Ibid.*, p. 36.

proportion may reach 18 per cent by 1970.[8] Such a trend would, of course, absorb some of the pressure toward increased local government expenditure for schools, particularly in urban areas. But the prospects are hard to anticipate and will be subject to many influences, including general economic developments, the ability of church groups to finance expansion of parochial schools, and the standards set in urban public schools.

The great unknown, of course, concerns such quality-related variables as rates of pay for public school personnel, the size of classes, special programs for particular pupil groups, and replacement of obsolete and mislocated public school facilities. I believe it can be argued from aggregative measures that some net gain has been made during recent years on all these counts. Teachers' salaries have risen more rapidly than living costs and also more rapidly than earnings in numerous other occupations. Public school classes average somewhat smaller than a decade ago. Construction of new public school classrooms has been considerably more than enough merely to provide for the net rise in total enrollment. Of the approximately 1.4 million public school classrooms in the country, something like 600,000 have been provided during the past decade,[9] so that the average age of the nation's public school plant now is undoubtedly much less than it was in 1950.

But the aggregative data tell only part of the story. They show nothing of the wide disparity which exists on all these scores between public schools in various areas--and particularly between major central cities and outlying fringe territory. Much of the most modern school plant, for example, is naturally located in the "growing edge" areas. For major cities an inventory by age of structure would undoubtedly disclose many examples of shocking antiquity and inadequacy. Dr. Conant has vividly described the array of pressing needs for better facilities and strengthened public school services in our metropolitan centers.[10] Large additional sums might reasonably be justified to meet these needs and to upgrade public schooling in metropolitan centers. But there are many institutional and financial barriers to such a development--including not only the competing pressures of other urban government services but also the discrimination that major cities commonly suffer, one way or another, in the distribution of state educational grants.

One moot factor involves the possibility of a general program of federal grants for public school purposes. However, the sums officially urged for this purpose would represent only a very minor part of all local school expenditure. The program submitted in 1962 to the Congress by the Administration called for grants totaling $2.1 billion during the following three years--or less than 4 per cent of the likely total of local public school expenditure during this period, aside from such federal assistance.

Even in the absence of a new program for federal school aid, we may probably expect that urban spending for public schools will continue upward more rapidly than urban expenditure for other purposes, but with some narrowing in the respective rates of increase.

Another and even more unpredictable aspect of future local spending for education concerns the college grades, as distinct from elementary

---

[8] Ibid., p. 36.

[9] Ibid., p. 37.

[10] James Bryant Conant, Slums and Suburbs, A Commentary on Schools in Metropolitan Areas (New York: McGraw-Hill Book Co., 1961).

and high schools. Most public education at the college level is now being provided through state institutions. However, in California there is provision for junior college education by local communities throughout the state, and New York City and a scattering of other communities in about half the states also have local public colleges or junior colleges. With expectations for a tremendous increase in college enrollment in the years ahead, some observers have urged that part of the load for public institutions be kept out of state residential colleges and universities by the development of community junior or four-year colleges. Widespread action of this kind could have a considerable impact on local public expenditure. In terms of the needs to be met, one might envisage local spending at the college level amounting to 10 per cent or more of the sums applied to elementary and high school education. Thus far, however, in terms of statewide totals, only California comes anywhere near this, and there are only two other instances--Kansas and Kentucky--where local public college spending equals 5 per cent of public school expenditure.

Probably we can anticipate a considerable spread of the public community college device, having a sharp impact upon local spending for education wherever it is developed. But the extent of this development will depend upon state and federal government underwriting of an important part of the costs involved. If such action occurs--and it would seem to make sense, if for no other reason than to handle through local institutions some of the tremendously increased load that otherwise will fall on state colleges and universities--education will surely continue to require a growing fraction of all local government expenditure.

Public Welfare

Our introductory analysis of "basic" urban government expenditure reported very little spending for this function. For most urban areas, this is a correct picture, since the public assistance programs that make up the bulk of all state-local welfare spending (and which have been arbitrarily excluded from the summary tabulation here, as already described), are predominantly financed from state and federal funds and, in most states also, are administered through state rather than local government agencies. However, in those instances where local participation in administration and financing does apply, the local contribution involved often represents a considerable fraction of all local government expenditure. Most widely, this situation involves primarily local responsibility for "general assistance"--provision for the needy other than through the federally aided programs for the aged, dependent children, blind, and disabled--but in a number of states local governments also administer and help finance one or more of the so-called "categorical" aid programs. Thus, while local governments provided in 1960 only about 12 per cent of all public assistance funds in the nation as a whole (with the federal government supplying 52 per cent and states the other 36 per cent), the local portion was 20 per cent or more in thirteen states--including, for example, Indiana, Kentucky, Michigan, Minnesota, New Jersey, New York, and Wisconsin.[11]

Even in such instances, the burden to local governments is a mere shadow of that to which they were exposed before the Great Depression of

[11] U.S. Department of Health, Education, and Welfare, op. cit. (footnote 6), p. 431.

the 1930's. A whole array of devices now limit the possible scale of local governments' responsibility for financial aid to the needy: the Old Age Survivors' and Disability Insurance system, currently providing benefits to about three-fourths of the aged population; state unemployment insurance programs; and the categorical aid programs, which care for the bulk of the assistance load, largely from federal and state funds. One observer has emphasized that the development of these arrangements has eased local fiscal tasks "in a manner, paradoxically, at once subtle and dramatic":

> ....Not many years ago, the cost of caring for the dependent aged and other needy persons constituted a charge on local budgets. The burden...had to be met, for the most part, from local property taxes. In periods of widespread joblessness, the load on local budgets became critical, often leading to total breakdown and the resort to all manner of fiscal expedients....
>
> ....The financial problems of the metropolitan areas are eased as these costly, and in many local cases unmanageable, difficulties are turned over to stronger government units.[12]

In part because of the maturing of the OASDI system, total spending for public welfare by state and local governments has risen far less rapidly in recent years than their expenditure for other purposes. State-local expenditure for public welfare in 1960 (including the portion financed from federal aid) was only 58 per cent more than in 1952, while the amount for education was up 125 per cent, and state-local expenditure for all other general government purposes together showed an increase of 86 per cent. It seems reasonable to expect a continuance of this tendency, with a relative further lightening that the demands of welfare spending place upon these governments.

However, in the particular states where public assistance arrangements still place a considerable load on local governments--and especially for "general assistance," which is so vulnerable to economic fluctuations --we may expect pressure for a shift toward more reliance on state government resources. Success in such efforts could provide some of the additional elbow room which urban local governments need to deal with other expenditure needs.

Health and Hospitals

About 5 per cent of urban local expenditure appears for this composite category. For local governments altogether in 1960, the Census Bureau reported the following amounts for health and hospitals:

		($ million)
Expenditure, total		1,899
Local public hospitals		1,463
Capital outlay	195	
Current operation	1,268	

[12]George W. Mitchell, "The Federal Impact on Metropolitan Finance," in Carl H. Chatters and others, Financing Metropolitan Government (Princeton: The Tax Institute, 1955), p. 145.

	($ million)
Payments to other hospitals	108
Health, other than hospitals	327
Revenue from charges for hospital services	650
State payments to local governments (including amounts from federal grants):	
For hospitals	95
For health	81

All these financial components have been increasing. Between 1952 and 1960 per capita urban local expenditure for health and hospitals rose at an annual rate of about 5.6 per cent. However, the rate of change has been slackening. In particular, hospital construction expenditure of local governments has not been growing in the past three years.

The bulk of state hospital expenditure is for mental institutions, where a marked gain in release rates has been achieved during the past decade, resulting in a significant reduction in total patient load relative to the nation's population. Aside from New York City and Washington, D.C., however, there are few local governments that operate mental institutions. Most local hospital spending is for general hospitals. Although these principally serve patients who are unable to pay in full for hospital care, revenue from charges equalled about one-half of all local governments' current spending for hospital purposes in 1960, and this fraction has shown some upward tendency toward recent years.

Sizable needs for more hospital facilities have been projected by the Public Health Service, but it seems doubtful--barring some major shift in institutional and financing arrangements--that local governments will have an important role on this score. The number of beds in local government hospitals is no greater now than twenty years ago. Most of the net increase in hospital beds during the past few years has been accounted for by private nonprofit hospitals.[13] Whether through the further expansion of hospital insurance coverage or a federally sponsored medical care plan, a further increase seems likely in the relative role of privately operated institutions, insofar as general hospital and nursing care are concerned.

On the other hand, a big increase in the health and hospital responsibilities of local governments will result if there is widespread application of proposals made by the Joint Commission on Mental Illness and Health. That group has urged that many of the persons who under existing arrangements would be cared for in large state mental institutions could be better treated in community clinics and hospitals equipped to deal with certain kinds of mental disability as well as with other illnesses. It has proposed state and federal programs to speed such a decentralization (both geographically and institutionally) in the care and hospitalization of the mentally ill.[14] Again in this instance the extent and pace of action will probably be limited by financing problems--and in particular by the inability or reluctance of local governments to handle some of the responsibilities now exercised largely by state governments.

---

[13]U.S. Department of Health, Education, and Welfare, op. cit. (footnote 6), p. 29.

[14]Joint Commission on Mental Illness and Health, Action for Mental Health (New York: Basic Books, Inc., 1960).

Altogether, the health and hospital expenditures of urban local governments can be expected to maintain about its present ranking among the various functional categories, and to continue to expand more rapidly than urban population. In the near future, however, a sharp growth in the scope of local public health and hospital activities seems unlikely, unless accompanied and induced by extensive federal-state financing of the extra costs locally incurred.

Urban Transportation

Streets and highways. In our introductory summary table we have estimated urban local spending for streets and highways in 1960 at $18.73 per capita, or about 9 per cent of all basic urban local government expenditure. These highway figures are probably less meaningful than those for most other functional categories. Above all, the limited focus here upon local government amounts involves a high degree of artificiality. Direct state spending for highway purposes is nearly double that of local governments. A sizable part of such state spending is for through-road facilities that are located within or directly serve urban areas, and the urban population also shares in the use of the remainder of the state highway systems. Furthermore, there is great interstate variety in the relative roles of local governments and the state with regard to the construction and maintenance of traffic arteries in urban areas.

Nonetheless, in the great majority of urban areas spending for street and highway purposes makes up a consequential portion of all local government expenditure. What are the prospects that this segment will increase or decrease, relative to other expenditure?

Looking at highway facilities as a whole, rather than merely those locally provided for urban areas, we can see strong pressures for continuing expansion. Numbers of vehicles and the mileage of highway travel have been expanding about 4 per cent annually, as against a population increase of less than 2 per cent. But the prevailing arrangements for highway finance are a constraint on the volume of highway work that can be accomplished, especially in periods of rising price levels. Most of the funds for state highways come from earmarked state and federal taxes on highway users, and these do not directly respond to inflation. The effect can be roughly measured by the fact that during the 1955-60 period, when other state tax revenue was rising more than 10 per cent annually, the yield of state vehicle-user taxes was increasing only about 6 per cent a year. During the past decade, a number of devices were used to deal with this situation--tax rate increases, large toll highway bond issues, other state borrowing for highway purposes, and a considerable increase in federal highway grants, with earmarking of revenue introduced at the federal level. The toll highway device has lost favor and there is always resistance to tax rate increases. So, if we assume some further upward trend in prices during the 1960's, it seems likely that highway expenditure of state and local governments will rise less rapidly than their spending for other purposes, and that this function will represent a somewhat smaller part of the state-local expenditure total.

A similar prospect may be suggested for urban local expenditure on streets and highways, but for a different reason. Only a very small part of such spending comes from local vehicle-user charges; most of it is financed from local property taxes, although state revenue sharing, special assessments, and borrowing to be later repaid from these other sources also enter into the picture. Thus, to be met locally, highway needs must compete directly with public schools and the many other services that are financed mainly from the property tax and other general revenue sources.

At least one and possibly two factors may tend to limit the highway demand on local governments during the coming decade. The most likely is the growing fraction of the total urban traffic load to be handled on facilities of the federal interstate system. Undoubtedly there are instances in which these facilities will actually increase the local burden by creating costly overloads and bottlenecks which local governments must remedy. But if one thinks of urban areas broadly as including closely settled fringe territory as well as core areas, it seems likely that the improved state highway facilities will handle much of the traffic for which counties and municipalities would otherwise need to make provision.

The other factor concerns the prospect for increased reliance upon public mass transportation facilities, as more fully discussed below. Smaller cities and towns are not likely to share in this development, but for at least some of our metropolitan centers a reversal of the trend toward ever-increasing reliance upon private car travel seems possible.

Airports and terminals. As indicated in Table 1, spending on airports and terminals represents about 2 per cent of urban government expenditure. Facilities for water transportation are operated by municipalities or special districts in only a few areas, but airports are publicly owned and operated in urban areas everywhere. These local government facilities generally are run on a self-sustaining basis, with capital outlays largely financed by borrowing plus federal grants, and charge revenue used to cover current operation and debt service. Following are local government aggregates for 1960, from the Census Bureau's summary of governmental finances:

	Airports ($ million)	Water transportation and terminals ($ million)
Total expenditure	316	157
Capital outlay	224	92
Current operation	92	65
Related revenues:		
Charges and rentals (gross)	140	104
Federal grants	35	--
State grants	24	1

Local government expenditure for airports has moved strongly upward during recent years, having nearly tripled between 1955 and 1960. It seems reasonable to anticipate that this trend will continue, to meet demands imposed by the increasing volume of air transportation. Such an expectation is strengthened by the fact that this local government function is largely self-sustaining, and thereby free from competition with facilities and services that must be financed from local taxes.

Urban mass transportation. One of the biggest unknowns of the coming decade involves the role of local governments with regard to public mass transportation. Transit service is now publicly provided by only about fifty city governments, but these include New York City and Detroit, as well as four other cities of over a half-million population. Special-district transit authorities serve the Boston and Chicago areas, and within the past few years transit authorities have also been set up in the San Francisco Bay Area and Los Angeles. Similar action is actively under consideration in several other major metropolitan areas. Although Philadelphia does not have public operation of transit facilities, it has instituted a system for city subsidy of private carriers to permit lower rates and prevent further reductions in the use of their facilities. Thus, there is already public provision or subsidy of mass transportation in each of the nation's five largest cities.

In 1960, publicly operated local transit systems had revenue of $581 million, and gross expenditure (including interest on utility debt) amounting to $750 million.[15] New York's subway operations accounted for a considerable part of these aggregates. Taken altogether, the other publicly operated transit systems broke even, with revenue of $370 million and expenditure of $364 million. But this was accompanied in many instances by a very low rate of capital outlay, and the use of outdated facilities which limit the attractiveness and popular acceptability of public transit service.

It seems most likely that local governments in the major metropolitan areas will become much more involved with mass transportation during the next few years--either through direct operation of facilities, leasing of publicly provided facilities for private operation, or subsidies in some other form for private operation. There is growing awareness of the illogic involved in financing urban highway facilities mainly from general tax resources while expecting a less expensive transportation medium to be financed entirely by its immediate users. However, inherited traditions as well as the competing demands of many other urban public services will surely delay and hamper widespread public action to bolster public carrier transportation. One major problem is to develop effective metropolitan instrumentalities that need not be bound by city boundaries in the scope of their operations. This requires state legislation creating or authorizing such large-area agencies, and providing methods for their financing. The federal government can be expected to stimulate and encourage local and state action on behalf of better mass transportation in major urban areas, but any funds it provides will probably represent only a limited portion of the total public financing necessary.

In this as in so many other areas of need for drastic change, it is difficult to avoid thinking in terms of the desirable rather than the possible or probable. Public provision of only a few dollars per capita annually--a minor fraction of present local spending on streets and highways--could probably have an important effect on standards and use of public transportation in our major metropolitan centers. How widely this takes place will depend on the solution of difficult problems of organization and financial arrangements for particular areas.

---

[15]All privately operated "local and suburban transit" operations (which excludes railroad commuter operations) are not much larger than this financially. Corporation income tax returns for 1958 indicated 1,082 private firms so engaged, with gross annual receipts of $716 million. Only 606 of this number showed a net income, altogether totaling $31 million.

Water Supply and Sewers

These two essential and traditional local government services accounted in 1960 for nearly one-eighth of all basic urban expenditure, as estimated above--water supply 7.2 per cent and sewers 4.7 per cent. During recent years urban spending per capita on these functions has been increasing nearly 6 per cent annually--somewhat more rapidly than the growth for most other functions except schools. This movement has been dominated by capital outlay, which has in recent years made up about two-thirds of sewer expenditure and about half of water supply expenditure. Aside from interest on related debt, 1960 Census Bureau figures regarding these functions were as follows:

	Water supply ($ million)	Sewers and sewage disposal ($ million)
Total expenditure	1,681	1,103
Capital outlay	343	767
Current operation	838	336
Related revenues:		
Charges for service	1,529	318
Federal grants	--	40

Future trends will depend mainly on what happens with regard to capital outlay, and this is likely to differ between the two functions, with some relative slowdown in water system outlays but a continued high rate of sewer system development.

The case concerning water supply may be briefly stated. Water-use estimates by the Business and Defense Services Administration (BDSA) have projected for the next two decades a rise in distribution by public water utilities (including the privately operated establishments that make up about one-sixth of the total) at a pace that only slightly exceeds the prospective growth rate for urban population.[16] In other words, the projections anticipate relatively little increase in per capita use of utility-supplied water. (During the past decade, on the other hand, there was a rise from 146 to over 170 gallons in average daily per capita use of utility-supplied water--calculated by reference to urban rather than total population.)

The BDSA projections apparently assume that self-supply by industrial water users will represent a somewhat increasing fraction of total water consumption. If this turns out to be correct, the pace of public water system construction could be stabilized--or at least drop back closer to the rate of growth in urban population.

Two other factors provide an extra stimulus, however, for a continued rise in expenditure for sewer and sewage disposal facilities: (1) growing concern for stream pollution, with federal and state government action being taken to expedite or force appropriate action toward better disposal practices by present sewer systems; and (2) the higher population density

[16]U.S. Bureau of the Census, Statistical Abstract of the United States, 1961, p. 166.

of urban fringe areas that were initially served by septic tanks but will increasingly need community sewer facilities.

Perhaps the nature of urban-fringe settlement during the next decade will have a significant influence also upon water system expenditure. During the 1950's, as we all know, there was a tremendous extension of "urbanization" that involved a relatively spread-out, shoestring and checkerboard development around major urban centers--a form of development that involves high costs per inhabitant for water supply facilities.[17] There are undoubtedly some areas where a subsequent population "filling in" process should require relatively little further water plant expansion. It remains to be seen, however, how much we shall have of such a filling-in process as against continued leapfrogging and sprawl in metropolitan-area development.

Police and Fire Protection

These protective functions accounted in 1960 for about 9 per cent of basic urban government expenditure--police protection 5.2 per cent and fire protection 3.9 per cent. In recent years local police expenditure has been increasing more rapidly than that for fire protection. On a per capita basis related to increasing urban population, the average annual rise for police has been 6.3 per cent, and the rate for fire protection has been about 4.1 per cent, which is less than the average for nonschool urban functions as a whole.

Salaries and wages make up the bulk of local government expenditure for these protective services--about seven-eighths of all policing costs and four-fifths of fire protection expenditure. Recent trends, therefore, principally reflect changes in the pay rates and numbers of police and fire personnel. In each instance, average earnings have been going up at about the same pace as those for nonschool local government personnel as a whole--about 4.3 per cent a year. But the growth in police personnel has been more rapid than that for fire protection forces, with the latter having only about kept pace with the 2.6 per cent a year increase in urban population. On the other hand, local police employment (on a full-time equivalent basis) has been moving up about 4.5 per cent annually.

A general continuance of these differential trends seems to be in prospect. Police force requirements are positively correlated with the population size of cities (i.e., large cities engage more police in relation to their population than do smaller ones), but fire force requirements appear relatively constant for cities of various sizes, at least down to those of 25,000 to 50,000. Historically, improved communication systems and other technological changes, such as automated systems for traffic control, have no doubt increased the effectiveness of each of these fields of urban service and have helped to hold down their rates of growth in cost. However, no striking new developments seem in prospect that would sharply alter their manpower demands in relation to urban population.

---

[17] There are 155 urbanized areas for which it is possible to compare 1950-60 changes in territory and population. Altogether, during this decade the population of these areas rose 30 per cent, but their territory increased 81 per cent, so that their average population density declined from 5,408 to 3,886 per square mile. For these central cities, average population density declined from 7,786 to 5,810 per square mile, while the average for fringe areas dropped from 3,167 to 2,617 per square mile. These figures have been calculated from U.S. Bureau of the Census, 1960 Census of Population, Vol. I, Part A, Characteristics of the Population: Number of Inhabitants, Table 22, pp. 1-40 to 1-99.

Thus, fire protection spending will probably move up somewhat less rapidly than urban expenditure for other nonschool functions, but police protection costs probably will continue to rise more rapidly than this.

Housing and Urban Renewal

Gross local government spending for this function shows less increase between 1952 and 1960 than local expenditure for any other category. In fact, in each of the intervening years 1953 through 1959, local expenditure for housing and urban renewal was actually less than in 1952. (The total amount dropped steadily from $766 million in 1952 to $435 million in 1956, and then moved upward to reach $612 million in 1959 and $850 million in 1960.)

A major portion of the sums involved pertain to low-rent projects constructed and operated by housing authorities, with initial capital costs financed by authority bonds carrying a federal guarantee; subsequent debt service is largely covered by federal contributions, with project rentals covering current housing-project costs. The balance of the functional class relates to local projects for redevelopment and urban renewal, which involve the acquisition, clearing or rehabilitation, and disposition of run-down property. Federal assistance is provided also for this program.

The scale of local government activity in these fields has been determined very largely by federal programs and policies--more so than any other aspect of local government. There was a drastic cutback in public housing construction between 1951 and 1956. (Not since 1953 has the number of new low-rent dwelling units started annually under local housing programs reached one-half the volume of such starts in 1951.)[18] This is a reflection of the relative defeat of the so-called "public housers" in Washington during the mid-50's. In turn, the expansion of local expenditure for housing and urban renewal since 1956 reflects somewhat less restrictive federal housing legislation, as well as progress of a growing number of urban renewal projects beyond the initial planning age.

Future prospects for this functional class of local expenditure are thus highly conjectural, since they depend so much upon the outcome of a sharp conflict, focussed mainly in Washington, between closely matched forces that have divergent views about the appropriate role of government with regard to housing and urban conditions. Dick Netzer recognized this uncertainty in his projection of state-local expenditures, by stating a possible range for this function (at 1957 price levels) all the way from $400 million to more than $1 billion in 1970.[19]

In any event, we must conclude that the eight-year period 1952 to 1960 provides no direct basis for projection. Especially through growth of urban renewal activity (as distinct from additional low-rent housing), it seems likely that gross local expenditure for this function will rise strongly during the next few years, and probably at a more rapid pace than local spending for purposes which more directly compete with one another for local tax resources.

---

[18] Housing and Home Finance Agency, Annual Report, 1960, p. 320.

[19] Dick Netzer, "Financial Needs and Resources Over the Next Decade: State and Local Governments," in National Bureau of Economic Research, Public Finances: Needs, Sources, and Utilization (New York, 1961).

Other Functions

Local government services not discussed above accounted for about 15 per cent of basic urban public expenditure in 1960. This remainder is made up as follows:[20]

Function	Per capita expenditure ($)	Per cent of all basic urban expenditure	Average per cent rise in per capita amount, 1952-60
Local parks and recreation.	6.31	3.0	8.6
Sanitation other than sewerage.	5.13	2.4	3.0
Local libraries.	2.12	1.0	6.2
General administration.	8.46	4.0	5.6
General public buildings.	2.30	1.1	5.4
Miscellaneous and unallocable.	6.91	3.3	10.4

There is little reason to assume any marked shift in the relative financial ranking of these components of urban expenditure. Each of the first three functions is strongly associated with urbanization, and on this ground one might expect them to be stimulated by growth in the average size of municipalities and urban areas. Moreover, recently developed urban fringe areas that have utilized private rather than public agencies for refuse collection and disposal are likely to move increasingly toward the latter kind of arrangement. On the other hand, certain factors may help to limit rates of growth in expenditure for these functions. One such factor is technological change. Improved equipment for street cleaning and refuse handling has undoubtedly increased the productivity of these local government activities during recent years; and further gains in productivity are likely. Also, the "general administration" category includes some large-volume clerical operations (for example, with respect to assessment, tax billing, and land records), which already benefit from automatic data processing in some large urban governments. Many more such applications seem in prospect. Similar technological gains may be expected for large library systems.

A more problematic factor concerns the recreation and parks function. Local expenditure for such purposes has been rising steeply, and with increased urbanization and higher standards of living a strong demand exists for additional recreation facilities. However, in the atomized local government situation which appears around many major centers, there is great difficulty in matching areas of benefit and financing. In our present motorcar age, particular fringe communities may refrain from developing park or recreation facilities lest they be swamped by "outsiders" from other parts of the total area. In some instances, we shall undoubtedly see an increased part of the cities' traditional concern for recreation in urban areas devolving upon counties or large-area special districts. On the other hand, the shift may be largely to state governments. In the latter event, much of the growth in public expenditure for parks and recreation would be reflected at the state level, rather than in the local government sector.

[20]The Census Bureau sources from which these relatively minor components have been drawn are based on sample surveys, so that the figures may differ somewhat from results that would have appeared from a complete enumeration.

SUMMARY

Urban local government expenditure in the United States has recently been rising some 10 per cent a year. On a per capita basis which takes account of the annual increase of about 2.7 per cent in urban population, the urban expenditure growth has averaged between 6 and 7 per cent annually since 1952. About half of this is traceable to price-level changes most directly expressed in salary rates of local government employees, while the balance presumably reflects growth in the range and per capita volume of urban government services and facilities.

Aside from a possible slackening in pay-rate changes, it seems reasonable to expect a general continuance of this over-all trend. Our review has suggested the possibility, however, that expenditure trends will differ materially among particular functions. We have anticipated a less-than-average rate of growth for local highway expenditure, water supply, public welfare, local fire protection, and, possibly, health and hospitals; a higher-than-average rate of increase in urban expenditure for education, sewerage, police protection, housing and urban renewal, and airports; and an indeterminate prospect for various other functions. However, barring widespread provision for public community colleges under local jurisdiction, or a significant federal school aid program, the rise in expenditure for education will not outrun growth in other urban expenditure as sharply as it has during the past decade.

The policies and programs of the federal government will have an important bearing upon developments in urban local expenditure--far more than might be suggested by the relatively nominal amounts of federal grant money that now goes directly to municipalities, housing authorities, and other urban governments.[21] This has already been illustrated in discussing public welfare expenditure.

Big catalytic possibilities can be seen also in federal action with regard to sewerage, housing and urban renewal, mass transportation, mental health, and community colleges. In most of these instances, state government policies also will have an important bearing--for example, in helping to determine whether or not community colleges or public community clinics and hospitals are actually developed to deal with needs now served mainly through state institutions, or in authorizing effective metropolitan agencies to deal with mass transportation. The states' role will be even more pervasive than all this might suggest, by their power to restrict or enhance the effective fiscal capacity of urban governments through legislation relating to assessment and taxation, debt limitations, state grants or revenue sharing, and local government structure.

In other words, the prospective pattern of urban public expenditure will be influenced not only by economic and demographic developments, but also by institutional arrangements which, in turn, in many instances may be determined at the state level rather than locally.

---

[21] The bulk of federal grants-in-aid are for highways and public assistance, involving payments to the states; of federal intergovernmental expenditures totaling $7 billion in 1960; only $642 million went directly to local governments, including $226 million for housing and urban renewal, and $245 million for (impacted) school aid.

3

Intergovernmental Aspects of Local Expenditure Decisions

by Selma J. Mushkin[1]

Many considerations enter into urban decisions for public services. Among these considerations is the "hard" tax dollar, against which are weighed: the budget of the previous year, cost-benefit comparisons at the margin, and the "Social-Problem-of-the-Year," carved out and dramatized in presidential and gubernatorial messages. Decisions about urban public services are influenced, too, by grants-in-aid from state and national governments. In this paper, we shall explore some of the issues raised by two questions about grants-in-aid: What restrictions do they impose upon local decisions? And, can grant programs be amended to achieve a more effective and efficient use of resources?

QUANTITATIVE IMPORTANCE OF GRANTS-IN-AID

Before taking up these questions, we shall briefly examine the nature of grants-in-aid and the amount of grant support of local expenditures.

The basic notion of the grant-in-aid is simply to provide a device for keeping program administration and decision making "close to the people" and to broaden the financial resources available for program support. A similar device is the shared tax which makes available to a community for expenditure, in accord with the decisions of local citizens, tax sources that can be administered more effectively and efficiently by a larger governmental unit. Use of superior taxing power for revenue raising purposes only, suggests a distribution of the proceeds of the tax back to the area in which it was paid. Use of the grant to further public programs requires a

---

[1]I am greatly indebted to my colleagues Norman Beckman and Jacob M. Jaffe for their suggestions and criticisms of an earlier draft of this paper. To Mr. Jaffe who spent many hours compiling the data on intergovernmental aids go my special thanks. The opinions expressed are the author's own and do not necessarily reflect the views of the Advisory Commission on Intergovernmental Relations.

distribution proportionate to some measure of program need. The rapid extension of grants during the depression of the '30's, when local property tax revenues dropped off sharply, largely accounts for the fusion of tax-sharing schemes with grants-in-aid. State sales and income tax levies were adopted at that time to meet the fiscal crisis; revenues from these levies were shared frequently with local communities as amounts earmarked for specific program expenditures. A part of the funds was distributed on the basis of need, a part in accord with the place of origin of the tax.

Aid from the state and national governments to the cities takes many forms, including, in addition to grants and shared taxes, loans and loan guarantees, technical assistance, personnel training and research support. The programs carried out directly by the state and the national government also vitally affect standards of living and economic development in the cities. And in terms of specific services, the urban family gets from a national agency its postal services, and from both national and state agencies it receives an assurance of safety of the drugs and food it buys. A complete listing would enumerate the items of expenditure of state and national governments and include that largest item of public outlay--national defense. The beneficiary groups for all the programs--local, state, and national--are families and the businesses that serve them. When the state provides unemployment insurance; or the national government, retirement benefits, the city's welfare rolls are reduced; and when the state locates a branch of a state university in a city the pressures on the city for higher educational facilities are relieved. These are examples of the complex intergovernmental expenditure impacts omitted from the present review.

The distribution of responsibility among levels of government for civilian public services attests to the success of the grant-in-aid and tax-sharing devices for intergovernmental co-operation. Local services are provided close to home with aid from state and national governments. The national government directly provides only a small part of civilian public services to the urban resident. In all, less than 8 per cent of all governmental purchases of goods and services are made in the course of administering national programs for civilian public services. Of the total state and local expenditures, localities account for almost two-thirds.

The amount of state aid to local government has increased over the past decade from about $4 billion in 1950 to over $9 billion in 1960, but state aid as a per cent of local general revenue has dropped off. While the figures cited include payment received from the state as reimbursement and sharing of the costs of services, they mainly represent grants-in-aid and share of taxes. About $7 billion of the $9 billion total was raised by the states through taxation, and more than $5 billion went to the local communities for schools. The remaining $2 billion total is approximate, since federal grants to a state become state monies and are not separately identified in the subsequent redistribution within the state.[2]

In addition to the indirect federal aids through the states, localities in 1960 received nearly $600 million directly from the national government

---

[2]Data in this paragraph are from U.S. Bureau of the Census, Historical Summary of Governmental Finances in the United States (Vol. IV, No. 3, of the 1957 Census of Governments); and Governmental Finances in 1960.

to aid housing and community development, airport construction, and waste treatment facilities, and to support education in federally affected areas. Direct grants from the national government to localities increased sharply between 1957 and 1960 and rose even more in 1961, as these figures from the Budget of the United States show:

Direct Federal Aids to Local Communities

Program	1961 ($ thousands)	1960 ($ thousands)	1957 ($ thousands)
Total	675,793	563,255	298,694
Urban renewal program	144,538	101,706	29,621
Urban planning assistance	3,045	2,554	651
Public Housing Administration	141,118	124,373	86,687
Airport construction	64,216	57,113	20,629
Waste treatment works	44,085	40,295	843
School construction in federally affected area .	71,042	70,553	67,068
School operation in federally affected area . . .	207,749	166,661	93,195

Additional grant programs have been adopted which would increase grants to cities, including area redevelopment aid, open space program, urban mass transportation, and low-income housing demonstrations, as well as additional project demonstration support under public health programs. Upward of a $30 million addition was estimated for 1962 for the first four of the programs enumerated.

Direct federal grants to local communities are also made as demonstration grants in support of projects. These project grants are made directly by the national government to a public or private agency which formulates an acceptable and approved project. At present, project grant authority is included in programs for hospital construction, air polution control, vocational rehabilitation, community health facilities, and others. While the funds are not large at present, they are increasing rapidly.

But despite the growth in national aids, both through the state and directly to cities, public services to the urban dweller, both those services which are an essential component of metropolitan living and more general types of public services such as education, are largely financed by local governments out of their own tax sources. The urban resident depends upon his local government for the cleanliness of his public eating places, the streets he uses, his water supply, emergency care in event of accident, police and fire protection, and for a goodly share of his recreational pursuits. In those places with the largest concentration of population, moreover, local governments tend to have a larger share of total program responsibility. In New York State for example, 75 per cent of direct general expenditures are made by localities; in California, 74 per cent; in Illinois and Ohio, 70 per cent; and in Massachusetts, 69 per cent. In seven of the ten states with population of 5 million or over--states that account for 40 per cent of the population of the nation--local shares of total expenditures exceed national average. (See Table 1.)

Grants to Jurisdictions within Metropolitan Areas

What share of metropolitan area expenditures is financed by intergovernmental aids?

To answer this question a sample was drawn of seventeen standard metropolitan areas, including the five most populous areas, the District of

Table 1. Local expenditures and revenues as a percentage of total state and local expenditures and revenues, 1960

States with 5 million population or over (arrayed by population size)	Local expenditures as per cent of state and local	Local general revenues as per cent of state and local	
		Levied by localities	Received by localities
United States	65.4	45.4	64.2
New York.	75.4	57.0	77.6
California	73.5	47.9	73.1
Pennsylvania	61.3	44.5	62.8
Illinois	70.3	54.0	67.3
Ohio	70.3	49.2	66.9
Texas.	62.3	43.9	58.7
Michigan.	66.7	43.9	67.5
New Jersey	77.2	64.0	72.8
Massachusetts	69.0	53.2	71.7
Florida.	64.4	44.6	61.5

Source: U.S. Bureau of the Census, Governmental Finances in 1960.

Columbia, and eleven others selected at random from a grouping of metropolitan areas by geographic location, and population size. Financial data from the 1957 Census of Governments, reporting intergovernmental aids by function, were tabulated specifically for this paper. In all, local general expenditures in the seventeen metropolitan areas amounted to $9.5 billion in 1957, or 53 per cent of all local general expenditures of the 174 standard metropolitan areas (as defined for the 1957 Census of Governments).

As Table 2 shows, the share of urban expenditures financed by state and national aids varies widely, both from function to function and from metropolitan area to area. In the New Britain-Bristol area 11.5 cents of $1 of local expenditures is financed by aid from national or state governments; in New Orleans the share of local expenditures aided is almost three times as large. Differences among governmental units in the share of expenditures met by grants within each metropolitan area are illustrated by the information as tabulated only in those areas that include portions of two or more states. However, within the New York metropolitan area, grants cover 73 per cent of welfare costs in the New York State portion of the area and 45 per cent of welfare costs in the New Jersey portion.

Grants per capita also have a wide range (see Table 3), from $15 per capita in the New Britain-Bristol area to almost $70 per capita in Los Angeles-Long Beach area. Excluding public welfare grants, the range narrows, but grants per capita are still more than three times as large in some areas as in others. Table 4 indicates the variation of per cent of state aids in the states in which these seventeen metropolitan areas are located. As the figures indicate, the major portion of state aids originates in state programs rather than as a counterpart of a federal grant program. Most of these state aids go toward support of such public services as are required by both the urban and rural resident, rather than in support of metropolitan type functions such as mass transit or slum clearance. In most states, state grants to local communities for education account for 50 per cent or more of total state aid for specific functions, and in some, such as Connecticut and Texas, 85 per cent or more of state aid goes for education.

Table 2. State and federal aid as per cent of local general expenditure, by function, for selected SMSA's, 1957

Standard Metropolitan Statistical Area	Total	Public welfare	Education	Highways	Health & hospitals	All other general govt. functions
SMSA's in which public assistance is primarily a local function:						
Los Angeles - Long Beach, Cal.	30.7	82.5	40.7	48.4	6.0	8.5
San Francisco - Oakland, Cal.	32.4	73.7	39.0	46.1	5.4	13.7
Santa Barbara, Cal.	30.1	88.2	28.4	96.8	3.1	8.5
District of Columbia:						
D. C. portion	18.2	45.3	1.8	12.4	0.2	30.2
Maryland portion	29.8	85.2	38.9	20.5	4.3	17.7
Virginia portion	20.8	73.6	30.7	23.5	4.3	6.4
Boston, Mass.	25.2	85.1	11.6	5.3	NA[2]	20.2
Minneapolis - St. Paul, Minn.	21.6	64.5	28.6	16.2	4.9	6.3
New York, N.E. New Jersey:						
New York portion	21.1	73.1	26.6	7.6	14.6	9.2
New Jersey portion[1]	11.5	44.8	16.4	21.7	5.9	2.9
SMSA's in which public assistance is primarily a state function:						
New Britain - Bristol, Conn.	11.5	38.5	19.0	2.6	--	3.8
Chicago:						
Illinois portion	13.8	103.5	15.4	29.5	7.4	1.0
Indiana portion	22.2	51.8	20.3	80.9	11.0	5.2
New Orleans, La.	33.9	13.4	61.0	34.7	7.8	21.5
Detroit, Mich.	29.3	30.7	37.7	59.2	15.8	14.8
Kalamazoo, Mich.	31.0	20.7	37.0	45.5	4.3	15.5
Toledo, Ohio	21.9	98.0	17.2	57.5	6.5	10.7
Philadelphia:						
Pennsylvania portion	13.8	9.7	23.3	27.7	3.5	8.8
New Jersey portion[1]	15.5	39.4	32.0	5.0	11.7	1.8
Providence - Pawtucket:						
Rhode Island portion	19.5	66.8	21.6	0.3	10.0	16.6
Massachusetts portion	19.0	38.3	15.0	2.3	NA[2]	20.8
Fort Worth, Texas	20.1	--	37.7	3.2	10.3	1.1
Lubbock, Texas	21.0	--	39.4	3.1	--	0.3

NA signifies data are not available.
[1]Public assistance administered in part by state and in part locally.
[2]Included in all other general expenditure.

Source: Compiled by Jacob M. Jaffe, in part from unpublished Census data and in part from U.S. Bureau of the Census, Local Government Finances in Standard Metropolitan Areas (Vol. III, No. 6 of the 1957 Census of Governments).

State-by-state differences in the allocation of functional responsibility and of taxation between localities and the state suggest wide variations in grant-in-aid patterns across state lines. Grants per capita correlate significantly with neither local ability nor local expenditures. Despite these differences there is an interesting cluster in the amounts received when these amounts are measured as the equivalents of effective property tax rates on full value of property. In ten of the seventeen metropolitan areas, grants are the equivalent of a 0.6-0.7 property tax rate.

Table 3. Per capita local revenue from state and federal governments, by function, for selected SMSA's, 1957

Standard Metropolitan Statistical Area	Total	Wel-fare	Educa-tion	High-ways	Health & hos-pitals	Housing & urban renewal	Other func-tions	General local support	Total less public welfare
SMSA's in which public assistance is primarily a local function:									
Los Angeles - Long Beach, Cal.	$69.73	$20.30	$34.67	$6.08	$0.67	$0.25	$4.49	$ 3.26	$49.43
San Francisco - Oakland, Cal.	69.03	18.67	33.69	6.03	0.97	0.48	6.09	3.11	50.36
Santa Barbara, Cal. . .	49.03	17.07	19.42	7.58	0.41	0.04	3.52	0.99	31.96
District of Columbia, total	37.82	3.75	14.74	2.06	0.19	0.88	1.98	14.22	34.07
D.C. portion (Federal only)	45.99	6.90	0.89	2.56	0.26	2.15	3.69	29.53	39.09
Maryland portion . .	38.79	1.87	26.87	1.97	0.19	--	0.79	7.10	36.92
Virginia portion. . .	25.01	1.72	18.65	1.47	0.09	0.22	1.11	1.75	23.29
Boston, Mass.	49.45	22.38	6.93	0.76	NA[2]	2.45	5.48	11.44	27.07
Minneapolis - St. Paul, Minn.	37.85	11.56	19.23	2.61	0.78	0.26	0.75	2.67	26.29
New York - Northwestern New Jersey, total	43.61	13.23	18.57	1.62	2.60	2.72	0.61	4.26	30.38
New York portion. .	52.54	17.00	21.18	1.42	3.26	3.10	0.77	5.80	35.54
New Jersey portion[1]	20.14	3.31	11.70	2.13	0.86	1.73	0.18	0.22	16.83
SMSA's in which public assistance is primarily a State function:									
New Britain - Bristol, Conn.	15.40	0.30	12.76	0.26	--	0.98	0.17	0.92	15.10
Chicago, total.	23.98	5.68	9.36	7.57	0.54	0.24	0.45	0.13	18.30
Illinois portion . . .	23.50	5.55	9.09	7.60	0.57	0.21	0.49	--	17.95
Indiana portion . . .	29.80	7.22	12.70	7.22	0.26	0.57	0.12	1.73	22.58
New Orleans, La. . . .	48.34	0.06	25.25	4.03	0.13	2.67	5.03	11.17	48.28
Detroit, Mich.	55.37	1.78	31.41	9.74	2.54	0.50	1.23	8.16	53.59
Kalamazoo, Mich. . . .	47.21	1.03	29.27	9.34	0.15	--	0.54	6.89	46.18
Toledo, Ohio	35.00	8.71	11.01	7.83	0.63	0.43	0.70	5.70	26.29
Philadelphia, total. . .	20.36	0.73	14.28	2.60	0.28	0.95	1.01	0.50	19.63
Pennsylvania portion.	19.54	0.34	13.56	2.61	0.19	1.05	1.18	0.61	19.20
New Jersey portion[1]	24.27	2.59	17.73	2.53	0.74	0.50	0.19	--	21.68
Providence - Pawtucket, total	22.65	4.03	9.96	0.06	0.26	1.62	0.37	6.35	18.62
Rhode Island portion	22.77	3.68	10.25	0.03	0.29	1.75	0.39	6.39	19.09
Massachusetts portion.	21.62	6.93	7.61	0.27	NA[2]	0.59	0.21	6.00	14.69
Fort Worth, Texas . .	24.05	--	22.72	0.37	0.47	0.25	0.24	--	24.05
Lubbock, Texas	25.23	--	24.60	0.53	--	0.06	0.03	--	25.23

NA signifies data are not available.

[1]Public assistance administered in part by state and in part locally.

[2]Included with "Other functions."

Source: Compiled by Jacob M. Jaffe from unpublished Census data. Per capitas are computed on the basis of population on April 1, 1960.

Table 4. State aids not originating in federal grant programs as a per cent of total state aid, by function, for selected states, 1957[1]

State	Functions not originating in federal grant programs			
	All functions	Education	Highways	Other
California	67	51	10	7
Connecticut	96	85	3	7
Illinois	89	44	34	11
Louisiana	86	77	8	2
Maryland	81	46	28	7
Massachusetts	41	22	8	12
Michigan	97	63	27	7
Minnesota	68	55	10	2
New Jersey	82	63	12	6
New York	73	56	6	11
Ohio	87	44	34	9
Pennsylvania	96	78	14	4
Rhode Island	86	57	29	--
Texas	96	92	4	0.4
Virginia	74	61	7	6

Note: Detail may not add to totals because of rounding.
[1]Those states in which seventeen selected standard metropolitan areas are located.
Source: Compiled from Bureau of the Census, State Payments to Local Governments (Vol. IV, No. 2 of the 1957 Census of Governments).

However, grants to New Orleans are the equivalent of a 2.6 per cent property tax rate, and grants to Chicago and New Britain, Connecticut, the equivalent of a 0.4 per cent property tax rate.[3]

Intra-State Distribution of Grants

What is the metropolitan area's share of all grants to local governments?

In 1957, 56 per cent of state and federal aid went to the standard metropolitan areas--areas with about 70 per cent of the nation's population (Table 5). The metropolitan areas received 55 per cent of state aids and 73 per cent of all direct federal grants to localities. Almost 85 per cent of public assistance grants to localities was received in metropolitan areas; in contrast only 32 per cent of health and hospital grants went to these areas.

Using the sample for seventeen metropolitan areas, the share of local expenditures supported by aid funds can be compared with the aid received in the rest of the state. Almost without exception, combined state and federal aid is a smaller share of local general expenditures in each of the seventeen metropolitan areas than in the rest of the state (Table 6). Grants finance a smaller part of school expenditures in the seventeen metropolitan areas than in the rest of the state except in the Maryland portion of the District of Columbia area, where federal grants to those school districts, which provide schooling for children whose parents work or reside on

[3]Property values estimated from 1957 Census of Governments, Taxable Property Values in the United States, by adjusting property assessments in accord with ratios of assessments to market value, for each county within the metropolitan area. Property values for some areas were estimated by applying the computed effective property tax rate of the central city to property tax collections in other portions of the metropolitan area.

Table 5. Local intergovernmental revenue from state and federal governments, by function, total, and in standard metropolitan areas, 1957

Functions	All local governments ($ million)	In standard metropolitan areas ($ million)	Per cent in SMSA's
State and federal revenue, total	7,539	4,229	56.1
From states	7,196	3,979	55.3
Direct from federal government	343	250	72.9
By function:[1]			
Public welfare	1,025	863	84.2
Education, total	4,266	2,136	50.1
From states	[2]4,119	2,038	49.5
Direct from federal	160	98	61.3
Highways	1,071	384	35.9
Health and hospitals	253	80	31.6
General local support	[3]640	413	64.5
All other	[4]284	[4]353	N.C.

N.C. signifies "not computed."

[1]The distribution for "All local governments" is based on state and federal payments (primarily for fiscal year ending June 30, 1957). The distribution for "standard metropolitan areas" is based on local revenue for fiscal years (often calendar years) that ended in 1957.

[2]Includes $25 million homestead exemption reimbursement to school districts: $12 million in Louisiana and $13 million in Iowa.

[3]Excludes $28 million homestead exemption reimbursement to school districts and special districts: $14 million in Louisiana, and $14 million in Iowa.

[4]Residuals reflect differences in timing and accounting between state and federal payments and local receipts. See footnote 1.

Source: U.S. Bureau of the Census.

federal installations, augment state grant funds. Grants for public welfare generally are higher in the seventeen metropolitan areas than in the rest of the state in which they are located. While state aid for public welfare programs, in almost all states with joint state-local programs, goes to local government within the state at a uniform matching ratio, the matching differs among the assistance programs and tends to result in larger grants to cities.

EFFECTS OF GRANTS ON LOCAL DECISIONS

While grants and related devices have put government "close to the people," they restrict the choices of a community by favoring the aided programs. What are these restrictions, and how does the grant device create them?

Grants as a Stimulus

Grants encourage communities to adopt new programs. The effectiveness of the categorical aid programs as a stimulus is implicit in some of the criticism of them. A hospital and public housing project are built, nursing home personnel are trained, children are vaccinated, a high school science laboratory is installed where before the community had taken no effective action.

Table 6. State and federal aid as per cent of local general expenditure, by function, for selected states, 1957

State and SMSA	Total		Public welfare		Education		Highways		Health & hospitals		All others	
	Selected SMSA's	Rest of state	Selected SMSA's	Rest of state	Selected SMSA's	Rest of state	Selected SMSA's	Rest of state	Selected SMSA's	Rest of state	Selected SMSA's	Rest of state
States in which public assistance is primarily a local function												
California	xx	37.8	xx	75.7	xx	48.6	xx	45.3	xx	9.5	xx	8.8
Los Angeles - Long Beach	30.7	xx	82.5	xx	40.7	xx	48.4	xx	6.0	xx	8.5	xx
San Francisco - Oakland	32.4	xx	73.7	xx	39.0	xx	46.1	xx	5.4	xx	13.7	xx
Santa Barbara	30.1	xx	85.2	xx	28.4	xx	96.8	xx	3.1	xx	8.5	xx
Indiana (Chicago - Indiana portion)	22.2	27.1	51.8	45.7	20.3	29.5	80.9	67.7	11.0	20.0	5.2	N.C.
Maryland (D.C. - Md. portion)	29.8	33.1	85.2	72.5	38.9	26.4	20.5	68.8	4.3	9.3	17.7	24.9
Massachusetts	xx	29.6	xx	30.0	xx	22.8	xx	22.9	xx	N.A.	xx	40.1
Boston	25.2	xx	85.1	xx	11.6	xx	5.3	xx	N.A.	xx	20.2	xx
Providence - Pawtucket (Mass. portion)	19.0	xx	38.3	xx	15.0	xx	2.3	xx	N.A.	xx	20.8	xx
Minnesota (Minneapolis - St. Paul)	21.6	31.2	64.5	30.8	28.6	36.6	16.2	20.1	4.9	25.5	6.3	16.3
New York (N.Y.-N.E. N.J. - N.Y. portion)	21.1	35.3	73.1	24.7	26.6	52.6	7.6	22.1	14.6	72.7	9.2	13.7
Virginia (D.C. - Va. portion)	20.8	32.1	73.6	75.4	30.7	41.4	23.5	18.1	4.3	16.0	6.4	13.9
States in which public assistance is primarily a state function												
Connecticut (New Britain - Bristol)	11.5	13.4	38.5	40.0	19.0	17.6	2.6	3.8	--	3.8	3.8	9.0
Illinois (Chicago - Ill. portion)	13.8	19.4	103.5	N.C.	15.4	22.7	29.5	58.3	7.4	33.9	1.0	N.C.
Louisiana (New Orleans)	33.9	53.1	13.4	--	61.0	73.8	34.7	24.8	7.8	27.8	21.5	23.3
Michigan	xx	43.8	xx	35.3	xx	53.0	xx	69.1	xx	10.7	xx	14.6
Detroit	29.3	xx	30.7	xx	37.7	xx	59.2	xx	15.8	xx	14.8	xx
Kalamazoo	31.0	xx	20.7	xx	37.0	xx	45.5	xx	4.3	xx	15.5	xx
Ohio (Toledo)	21.9	27.3	78.0	62.9	17.2	23.7	57.5	59.9	6.5	4.8	10.7	16.0
Pennsylvania (Philadelphia - Pa. portion)	13.8	31.4	9.7	4.6	23.3	60.0	27.7	50.8	3.5	64.5	8.8	N.C.
Rhode Island (Providence - Pawtucket - R.I. port.)	19.5	13.8	66.8	20.0	21.6	7.8	0.3	8.8	10.0	--	16.6	27.0
Texas	xx	23.5	xx	--	xx	43.6	xx	7.7	xx	2.1	xx	1.0
Fort Worth	20.1	xx	--	xx	37.7	xx	3.2	xx	10.3	xx	1.1	xx
Lubbock	21.0	xx	--	xx	39.4	xx	3.1	xx	--	xx	0.3	xx
New Jersey[1]	xx	17.7	xx	54.6	xx	29.3	xx	25.9	xx	7.3	xx	N.C.
N.Y. - N.E. N.J. (N.J. portion)	11.5	xx	44.8	xx	16.4	xx	21.7	xx	5.9	xx	2.9	xx
Philadelphia (N.J. portion)	15.5	xx	39.4	xx	32.0	xx	5.0	xx	11.7	xx	1.8	xx

N.C. signifies "not computed."

[1]Public assistance administered in part by State and in part locally.

Source: Compiled by Jacob M. Jaffe, in part from unpublished Census data and in part from U.S. Bureau of the Census, State Payments to Local Governments (Vol. IV, No. 2 of the 1957 Census of Governments) and Local Government Finances in Standard Metropolitan Areas (Vol. III, No. 6 of the 1957 Census of Governments).

The effectiveness of the stimulus may be partially explained by the way programs develop. The source of a nationwide program lies frequently in a community experiment. A community tries experimentally--under private or public auspices--to meet a problem, to relieve the burdens of hospital costs by providing organized home services, to reduce delinquency and crime by restoring a neighborhood, to alleviate the shortage of jobs by encouraging new industries. If the problem is sufficiently widespread and the experiment is reasonably successful, special interest groups will advocate more experimentation and more governmental programs to implement the activity on a broader scale. The citizens whose interests are immediately served and the groups with which they align themselves turn to the cities, states, and national government for action. Once a proposal becomes part of the President's or Governor's legislative program, public attention is focused sharply on the issues. Legislative study and debate further intensify interest, both pro and con, and pressures for local action gain added momentum.

Once a national or state aid program is adopted, the offer of financial aid is added to other pressures on behalf of a specific program. The restriction in choice of the local government originates in the added push given a program as a consequence of a grant offering. It would be a mistake to conclude that in all instances the offerings favor the adoption of a program and the acceptance of the standards set as a precondition for the grant. Many communities and some states do not accept all offerings.

We can look at this problem from a different angle. As a result of the grant offerings, a given amount of expenditures can be had in a community at different costs to the local taxpayers. At one extreme, for example, the building of a new fire house or police station will require 100 per cent local funds; at the other, an urban relocation project will require no local tax support. A rational choice, other things being equal, would suggest adoption of aided programs, and among the aided programs those with the most favorable (lower local) matching.

Grant Standards

Grant standards restrict the choices of a local government in formulating the specific content of the program and its administration, by encouraging them to qualify for aid and to conform to the standards, requirements, and policy directives of the aiding government.

Among the restrictions there may be, for example, a requirement that the administrative agency must be a health or an education agency in the community. Salaries must be "not more than," or, alternately, "not less than," for a specific type of employment, and so forth. The most important types of restrictions are essentially of three types: those affecting eligibility for public services, those affecting the qualifications of public employees, and those affecting amounts of expenditure per public beneficiary.

Restrictions of this type on local decisions stem from the nation's or state's objectives to help assure effective administration and quality of services, despite program decentralization. In our federal system of government, a sharp distinction must be made between the standards attached to the grant offered by the national government, and those attached to state aids.

Standards in federal aids are a way of directing the specific content of a program in the states or localities. They are a way to encourage adoption of a program in which a "workable program" is formulated, including revised zoning ordinances, land use plans, and plans for community facilities. The prerequisites for eligibility are made uniform throughout the state, tests of residency are broad enough to approach nationwide coverage, and merit systems are established for hiring and retaining employees. These specific standards alter the choice before a state or community. It can either accept the package deal, funds and standards, or go it alone without the program, or with a self-financed program.

The strings attached to state aids, however, are not always essential to the state's enforcement of standards in the community. The state has the authority to establish statewide eligibility, to certify the qualifications of local employees, to determine the specific content of local programs, and to do so without using financial inducements. Except in those instances in which the state's constitution provides to the contrary, the powers of the local governments--including powers of taxation, borrowing, and expenditure--are derived from the state. These differences between the national government's relations to states, and the state's authority over local governments are reflected in differences in operation of that single instrument of intergovernmental co-operation: the grant-in-aid. Conditions and standards are an integral feature of the federal grants, while standards are usually not combined with state aid programs. Nevertheless, if a state calls for a local activity which requires large local expenditures (for example, public schooling for all children within the state) the state necessarily provides aid to localities to implement its policy.

Two programs, both having children as their immediate beneficiary group, may be used to illustrate the differences in underlying purposes of grant provisions. In many instances, an important objective of state aid to education has been to strengthen decentralized control of education; whereas a primary objective of federal-state public assistance grants has been to obtain uniformity of aid to dependent children within the states.

Uniformity among local jurisdictions in the administration of the Aid-to-Dependent-Children program, and in eligibility for assistance, is due to the standards required of state assistance plans by the Social Security Act. To be approved, a state assistance plan must be statewide in operation, and the state is required to administer the plan or supervise its administration. The Social Security Act further limits the time an applicant must be a resident of a state in order for him to be eligible for assistance; it also defines a "dependent child" and, except for medical care, requires that the payments be unrestricted money payments. The definition of "need" and the amount of assistance to be given, however, are left to the states, with the proviso of statewide operation.

These standards were intended to remedy the situation prevailing when the Social Security Act was enacted. While authority for assistance to dependent children in the form of aid to their mothers, or pensions, existed in almost all the states before the federal statute was enacted, aid was actually granted in less than half the counties or other local governments under the permissive statutes of the states, and costs in all but a few were borne entirely by the county or township.

The conditions of federal grant participation resulted in a marked change in the finances, content, and administrative structure of the aid program. In twenty-six states, the program is administered directly by the state government without local financial participation. The largest sharing of costs by localities occurs in New Jersey where the local share is 28 per cent of the costs of the Aid-to-Dependent-Children program, followed by New York State with its local share of 26 per cent.

State restrictions on local communities, imposed in conformity with national requirements, were recently dramatized by the Newburgh "case." Newburgh's denial of assistance, and its imposition of more restrictive conditions than those prevailing throughout the state, led to state action. Had the state not intervened, the federal government would have had only that ultimate weapon: cutting off $65 million federal aid to dependent children in the state--almost half the funds for the program.

Responsibility for schools rests on the state, and state constitutions provide in one form or another "for an efficient system of common schools throughout the state." Tax and other powers have been delegated to local boards of education and local communities for this purpose and grants supplement and broaden local taxing powers. Only in two states—Delaware and North Carolina—are the school programs essentially state administered and financed. A battery of state aid programs has been developed to help assure local control of education but, at the same time, to facilitate the achievement of a minimum level of education throughout the state. More recently a number of categorical aids have been adopted by the states to encourage the establishment of special educational services such as driver training, classes for handicapped children, and use of audio-visual aids. These special grants account for only a small part of the $5.3 billion state aids to public schools. For the most part the monies go to localities as grants per child, as equalization aids, or some composite form tied to tax sharing. Flat grants are made without restriction and the strings for the equalization aids are largely minimum tax effort rather than program requirements.

Grant Adjustments

Local communities, after adopting an aided program and accepting its conditions, face a set of altered circumstances. Should the state reduce its aid, they must decide whether to curtail a public service with its current beneficiaries and administrative arrangements. Economic, technological, and social developments in a community—even if the absolute amount of state aid, or even the relative state sharing of cost is not changed by law--may so recast its need for services and the size of its beneficiary group, as to change drastically the local undertaking.

It is one thing for a community to consider an escalator provision in state aid before it adopts a project; it is another when the changes in aid are not known ahead of time. Patently, a downward escalation of shares of cost financed by grants provides less weight in favor of action on the aided programs than on programs that call for continuity or enlarged aid. Too little is known about the effect on local decisions of escalator provisions, but there are some indications that communities do take account of them in accepting program responsibilities. For one thing, experience under the demonstration project provisions of the vocational rehabilitation program suggests this conclusion.

Inflexibility of Formulas

Inflexibilities in aid formulas, and the lack of automatic stabilizers of state support in relation to expenditures, create delays that throw the burden of added financing on local communities.

The inflexibilities in state aids are of many kinds:

1. Program costs which the state will share are set when an aid program is enacted. Changes in prices, salaries, and in methods of operation occur that make the original levels unsuitable. As a corrective, automatic price or cost per unit adjustments have been proposed and, in some instances, applied. But the main remedy is still to amend basic legislation. It is not a simple matter, however, to build into legislative authority automatic provisions to vary amounts of aid. Apart from the technical problem of finding suitable indexes that will take account of the complex factors affecting program costs (for example, the effect of a shift from use of arsenic to use of antibiotics on the cost of a venereal disease program), these automatic devices raise squarely the issue of legislative control of public expenditures.

2. Beneficiary loads change with changes in employment, income, population movements, and advances in science. Medical advances, for example, can reduce a tuberculosis caseload and make a sanatorium an outmoded institution. Those grant programs that have fixed dollar sums to allocate among participating governmental units are not responsive to changes in the size of direct beneficiary groups. And even those whose aid varies with numbers of persons receiving the service do not gain the flexibility required. Movement of population from central city to suburbs, and the concentration of low-income groups in the cities, resulting in a backlog of neglected education and health, can so alter the quality of public schooling or public health services as to reduce the use of public services by others remaining in the central city. Distribution of aid on the basis of current number of users, in effect, builds a box around the program and prevents the enlargements of public expenditures that are required to meet both the community's current need and the backlog of neglect.

3. Various "floor" provisions in grant programs, tied to local effort (actual or hypothetical), are often based on fixed mill rates on the value of property in a locality. Property values, however, frequently are assessed valuations, and sometimes even are set at a base-year level. Lags in property reassessment and inequalities in assessment tend to defeat the original legislative objectives of these provisions.

In the case of grants for public schools, the combination of statutory foundation program costs and state sharing of these costs in excess of fixed mill rates on property, under present circumstances, creates a sizable downward push on state aid. Costs of schooling per pupil are increasing and property valuations in urban communities also are rising. Statutory formulas with rigid limits cannot adjust automatically to altered circumstances. New state-aid legislation has been enacted, but with it have been delays and debate about relative shares for the different communities. During the past decade localities have had the choice in any year of lowering the quality of education, or increasing property taxes. (The recent decrease in the ratio of state aids to total local revenue reflects primarily the lack of adjustment of state grants for education.)

A number of forces are helping to counteract the early emphasis of state aids on public services in rural areas. Educators have encouraged the use of foundation program supports in which requirements for expenditures are determined with reference to detailed budget items. These separate the costs of instructional supplies, school maintenance, and teachers' salaries--and salaries are determined on a schedule allowing higher compensation for teachers with advanced training and longer experience. This type of equalization formula raises the level of possible support for those districts that can attract teachers with more experience and training.

Special aids have been introduced in some states for the purpose of meeting the educational problems in cities and suburbs. Some states introduced new grants to help finance construction of schools in municipalities. New York State has special programs of aid for the education of non-English speaking pupils, culturally disadvantaged children, and also aid for educational services to children with special behavior problems. Also, proposals have been advanced for modification of equalization aids not only to adjust amounts of the foundation program, but also, in some instances, to move toward an open-end grant. Under such an open-end grant (in effect in a few states) matching shares for the state and local governments would be set by the relative capacity position of each district but would extend over the whole amount of the local expenditure per pupil as a stimulus to expanded public school services.

Reforms of property taxation have gained momentum, and accordingly provide a more adequate measure of the relative capacity of rural areas, cities, and suburbs. Patently, discrepancies among jurisdictions in assessment practices defeat the objectives of equalization grants for schools. Recognizing this, twenty-five states provide for the use of equalized property assessments in local jurisdictions as the basis for local school support; seven other states have turned to some index of capacity unrelated to property values because of the lack of reasonably comparable information.

Use of equalized property assessments as a base for educational grants from states to localities has frequently provided the impetus for property tax reform. In many states, property tax equalization is undertaken principally as an adjunct to state grant policy, without changing, except by the force of fact and publicity, local assessment rolls for local tax purposes. State action on equalization of property assessment stemmed from the apparent inequity of rewarding with enlarged grants those jurisdictions with comparatively low assessments, or encouraging them to compete with other local governments for under-evaluations.

There are some indications that the property values in cities have not increased as much as property values in other places in the state.[4] Substitution of current property-value data for base-year property assessments would help to raise levels of state aids for urban schools.

In summary, the present grant programs alter local decision in favor of introducing a new program, and in favor of accepting the specific standards on which the aid is conditioned. Their inflexibility and slow response throw a good share of cost increases--brought about by economic and

---

[4]New York City, Cook County, Los Angeles, and San Francisco rank below average in per capita market value of property in an array of counties within the state in which these cities are located.

demographic changes--on locally collected taxes, principally on the property tax. The voters' reactions to additional local tax levies largely determine the level of local expenditures.

GRANTS AND THE EFFECTIVE USE OF RESOURCES

Some tools have been developed for more efficient structuring of local expenditures and more effective choice in the allocation of resources. The few that are discussed in the following sections mainly concern the use of grants to improve the structure of local governments and to encourage the application of such guides to local decision making as planning, regional economic accounting, and cost-benefit analysis. Out of this discussion it becomes clear that grants can be offered in such a way as to improve the basis on which local choices among categories of goods and services are made.

It has become conventional to challenge the use of categorical grants to local communities because they restrict the choices of the local community in distributing funds among competing uses. The present categorical grant programs, however, on analysis appear to work in much the same way as a block grant, namely, expenditure decisions rest on local tax decisions for the most part. Neither block grants nor present categorical aids result in "proper" allocation of resources when benefits accrue to outsiders, or taxes used inside the cities are shifted outside. Only by removing from the local taxpayer the burden of financing benefits to outsiders, either by new payment devices such as a commuters benefit tax, or by grant subsidization in proportion to the cost of services to outsiders, can local decisions yield an allocation of resources which is effective and efficient from a state-wide or nation-wide perspective. The analysis presented, furthermore, suggests that certain conditions or standards imposed by the granting agencies, national or state, would facilitate rational allocation decisions. Standards such as that calling for interagency co-operation in planning and program development are illustrated.

Grants and Structure of Local Governments

The technical and engineering aspects of important public services are such that these services can only be provided on a regionwide basis, or the cost structure is such as to make for large economies of scale. It is not technically possible to curb air pollution in one section of a region, to clean up the water on one side of a stream, to build a useful highway that has no access roads. While a piecemeal attack on slums is a technical possibility, and zoning and housing restrictions can be technically imposed for a limited area, the consequences are a tidal wave elsewhere in the region. Inter-local governmental co-operation or consolidation is frequently a requisite for efficient use of public facilities and professional personnel. Grants to a local jurisdiction can keep a small school or health district going, even though operations are costly and ineffective, or grants can be designed to encourage inter-local jurisdictional co-operation or consolidation of administrative units.

The Advisory Commission on Intergovernmental Relations summarized the problem of fragmentation of governmental units in this way:

"The local government pattern in metropolitan areas is unbelievably complex. At the time of the 1957 Census of Governments, when 174 standard metropolitan statistical areas had been designated, a total of 15,658 separate local governments were identified in such areas: 266 counties, 3,422 municipalities, 2,317 townships, 9,185 independent school districts, and 3,180 other special purpose districts. This indicates an average of about ninety local governments per metropolitan area, but there is a range from a few units in some instances up to several hundred in some metropolitan areas. As designated in 1957, the Chicago-northwestern Indiana area had 954 local governments, and the thirteen counties making up the New York-northeastern New Jersey complex had 1,074."[5]

While historically grants to small school districts encouraged their existence, more recently states have used grants to encourage school redistricting and consolidation. In a recent discussion of the impact of state financial aid on school district structure, Leslie L. Chisholm, professor of educational administration at the University of Nebraska, said: "Provisions in a State school finance program for aiding local districts in their capital outlay expenditures may be one of the strongest features encouraging school district reorganization."

State aid for pupil transportation also stimulates reorganization. However, as Professor Chisholm emphasized, unless small districts are denied flat grants and equalization aid is based on a "careful consideration of the need in a properly reorganized school district," ineffective educational districts will be continued.[6]

Changes in the structure of administration of other public services also have been stimulated as an adjunct to grants. For example, multi-county or other inter-local governmental administration has been encouraged by the federal library grant program. State plans of the pooling of library services on an inter-local governmental basis, while not made mandatory by the federal act, have been stimulated by demonstration programs supported under this act and by professional association standards. Similarly, the pooling of personnel and funds for multi-county public health centers, and consolidation of city-county health departments have been encouraged under the Public Health Service grant programs.

Grants and Guides to Local Decisions

A number of guides have been developed to help urban policy makers and voters make rational decisions on new or changing programs. Grant programs in some instances encourage use of these guides; in others, grant provisions can be modified to serve this purpose, and in still others the grant arrangements create barriers to their effective use.

Among the guides to public policy formulation are comprehensive plans and the planning process, regional economic and social accounting, and cost-benefit analysis of programs. The questions we are asking here are: How do the grant-in-aid programs fit into the scheme of these guides?

---

[5] Advisory Commission on Intergovernmental Relations, Government Structure, Organization and Planning in Metropolitan Areas (Washington: U.S. Government Printing Office, 1961), p. 13.

[6] "Excerpts from Regional Conference Highlights," in Financing Education for our Changing Population (Washington: National Education Association, April, 1961).

Do the grants encourage, facilitate, or deter the use of guides to efficient and effective use of resources for local public services?

A grand scheme of public service requirements, of the costs of these services, of their interaction and their relation to the private market place has not been formulated. Not only would a computational system that serves as a base for planning and program evaluation have to include a multitude of variables, but for each a predicted value would be required to fit the system into time. At a minimum, predictions would have to be made sufficiently in advance to allow public officials to choose, not in terms of what was, but in terms of what is and of what will be.

Planning grants. Planning for the physical development of an area has long been recognized as an important instrument of effective local government. Many aid programs provide for the development of advanced plans for specific types of facilities, such as public housing, hospitals, and slum clearance. In some cases, federal aid has been conditioned on the development of long-range plans. To the battery of federal grants for community project planning, procedures have recently been added for combined highway and general urban planning grants. Rapid population growth in metropolitan areas has sharpened the need for comprehensive planning and for co-ordination of area wide functions, if not consolidation of these functions in a single administrative unit.

Barriers to effective and co-ordinated planning in metropolitan areas arise from both the fragmentation of aids for specific functions and fragmentation of governmental authority. Concern of federal, state, and local groups with this fragmentation is reflected in the priority assigned by the Advisory Commission on Intergovernmental Relations to proposals that should help remove these barriers. The Advisory Commission has recommended, among other things:

1. Enactment of state legislation to authorize two or more local units of government within a metropolitan area to exercise jointly or co-operatively any power possessed by one or more of the units, and to contract with one another for the rendering of governmental services.

2. Enactment of state and national legislation to require review by a metropolitan planning agency of applications for federal grants for construction projects from political subdivisions within the metropolitan area, (airport construction, waste treatment works, urban renewal, public housing, hospitals, and urban highways), without necessarily requiring that such an agency have the power to approve or veto.

3. Enactment of a new program of federal financial assistance and provision of federal and state technical assistance to metropolitan area planning agencies on a continuing or support basis.

4. Advance approval by the Congress of compacts creating interstate planning agencies.

The Advisory Commission emphasized: "To be worthwhile to serve a useful rather than an academic purpose, the respective facets of metropolitan area planning must be closely geared into the practical decision making process regarding land use, tax levies, public works, transportation, welfare programs, and the like."

Accordingly, one approach toward the solution of the problems created by categorical grants to different governmental units within a metropolitan region is to encourage the formation of metropolitan planning agencies (for regional and specific construction proposals) with access to the public through publication of their findings. Many would extend the Advisory Commission's proposals from federally-aided to state-aided construction projects, and would strengthen the authority of the planning agencies.[7]

Regional economic accounting. In addition to the engineering ideas for a metropolitan area, which describe zoning patterns, transportation routes, water and sewerage lines, parks and playgrounds, planning also rests on a mapped flow of economic activity in and out of the metropolitan area and of the activity within it. Regional economic accounts may be viewed as a mapping of these flows and a description of their magnitudes. For the policy maker, these accounts help to portray the patterns of economic life in the community and describe the relationship of local government to private businesses and households.

Moreover, they provide the basis for predicting the role of the area in the economic development of the nation. If the national markets of the major industries of an area are declining, if plants are automating their operations, or if a substantial industrial relocation is indicated, an engineering plan for public and private facilities may be obsolete shortly after it is off the drafting board. Lesser shifts, such as outward movements of shops and services, can also relegate an engineering plan to the scrap heap.

Decisions in metropolitan areas depend on economic accounts that give full recognition to the human resources and their productivity potential as one component of economic development. Harvey Perloff has emphasized the need for data on human resources as a basis for metropolitan area decisions.[8] He has contributed significantly to a recognition of services to people as a component of planning and as investment in human capital, which can contribute to higher productivity and higher income levels in the future and to a decline in public dependency. The notion that city expenditures can add to the wealth and income resources of a community is not new, however. A number of studies in the early 1900's estimated the public health expenditures required in the cities and evaluated these expenditures in terms of the dollar value of lives that would be saved.[9]

Adequate economic accounting for long-range planning, including planning to improve the quality of people, however, requires skilled professional assistance not always available even to large metropolitan areas

---

[7]Intergovernmental Relations Subcommittee, Government in Metropolitan Areas, Commentaries on a Report by the Advisory Commission on Intergovernmental Relations, House of Representatives, 87th Congress, 1st Session, December, 1961.

[8]Harvey S. Perloff, "Relative Regional Economic Growth: An Approach to Regional Accounts," in Design of Regional Accounts, Werner Hochwald, ed. (Baltimore: The Johns Hopkins Press for Resources for the Future, 1961), pp. 38-66.

[9]Quoted in Irving Fisher, "Report on National Vitality, Its Wastes and Conservation," Bulletin of the Committee of One Hundred on National Health (Washington: U.S. Government Printing Office, 1909), pp. 121-23.

and less so to the smaller jurisdictions within these areas. Grant standards can be extended to encourage planning that rests on a foundation of economic accounting. The area redevelopment program in fact conditions federal aids on such economic studies. However, technical assistance and adequate staffing of state and federal agencies for the purpose of making economic studies are the chief prerequisites.

Cost-benefit analysis. Still another guide for local decisions is the comparison of benefits from a public undertaking with their cost--an analysis, that is, of the resources of men and materials required for the output of a public service, on one hand, and of the utilities provided on the other. Cost-benefit analysis essentially represents a way to apply a private market calculation in public decisions, substituting, let us say, a maximization of social benefits for a maximization of profits.[10]

Conceptually, matching costs and benefits at the margin over the broad range of public and private goods and services yields a test of efficiency in the allocation of resources. If we had a shopping center where the consumer could purchase his public services at a price, relative consumer preferences for public and private goods would be measured by the usual market tests and the problem of resource allocation among classes of services would be determined by consumers. This image of a shopping center for public services, however, except for the emphasis it gives to use of fees and prices in financing wherever appropriate, does not ease the task of the public policy maker. Most local services are provided by the government rather than by private firms because the price system would work too harshly as a rationing device, or because private purchases--where there are large external benefits to others in the community--will lead to misallocation of resources, or because the service can be had only collectively for the community as a whole. An alternative to the private market test is the systematic evaluation of benefits and costs of competing public programs as guides to those who are responsible for the formulation of public policy. What the policy makers and voters in a community need to know to guide their choices is whether an additional expenditure on one public service will yield a greater or lesser gain than an additional expenditure on other things. Cost-benefit analysis lends itself more readily to evaluation of choices among construction projects than to valuations of other public program choices. For one thing, the relative benefits of service programs are more difficult to quantify than are the outputs of specific construction projects such as water resource development, or highways.

Grants alter the over-all framework of a benefit-cost analysis, and measures of the benefit and cost flows in and out of a community tend to point up requirements for grant support.

For the local policy maker, cost is translated into local tax rates (with an eye, however, on the taxes paid inside the locality to other jurisdictions). Benefits are the value of services to the local voter. But taxes are shifted and people and businesses move. Families work and shop outside of the community of their residence and people and goods move from area to area and across state lines. Even a partial listing of the flows of tax payments and benefits would include the following:

---

[10]For a fuller discussion of the techniques of cost-benefit analysis, see the paper by Nathaniel Lichfield and Julius Margolis in this volume.

1. Benefits received by outsiders from services provided in a community:

   (a) in the course of daily traffic flows (e.g., mass transit, highways, etc.);

   (b) by periodic traffic flows (e.g., use of public libraries);

   (c) by change in residence or employment of persons who have been the recipients of large amounts of public services (e.g., education, health);

   (d) essentially unrestricted to geographic limits (e.g., research findings).

2. Benefits accruing to a community from public services provided outside:

   (a) in-migrations of trained and professional manpower;

   (b) technological and scientific advances.

3. Taxes levied by a community, paid by outsiders.

4. Taxes paid a community, levied outside:

   (a) state taxes;

   (b) taxes of other communities.

If we think of costs in terms of the local budget, as additional taxes to be raised locally, and of benefits in terms of the value of services to voters and their families in the area, then a rational set of decisions would suggest more spending if taxes can be shifted outside and less if benefits accrue in substantial share outside.

When taxes are shifted and benefits spill over from one political jurisdiction to another within a metropolitan area, the effect on the decision process is under-allocation of resources (relative to supply) for those services used extensively by commuters. The resident-consumer-voter can be expected to be willing to pay only for the quantity and quality he wishes to purchase for his own and his family's use. This line of reasoning offers one plausible explanation for the frequent observation that municipal services are under-supplied. It suggests, as a corrective, outside grant support of those services whose benefits accrue to outsiders. It also indicates compensatory return in grants in those instances in which the taxes paid inside go out, or in which negative benefits (expenditure loads) flow in.

While the flows have not been measured over the broad range of public services, public action is taken in these terms. Gasoline and other motor vehicle levies, for example, are frequently shared with localities for highway use in recognition of the place of origin of the tax payments. Concern about the flows is reflected in the familiar question: How much is paid in this city and how much do we get back from the state? It is evident, too, in the continuing conflicts between upstate and downstate.

It is not only the location of tax payments that is at issue, but the situs of public expenditures. On an interstate basis, population movements result in a welfare cost in Chicago originating in low levels of education in Mississippi, or in a public hospital charge in New York as a result of inadequate health services in Puerto Rico. These interstate movements provide part of the rationale for national grants to help assure some minimum public services throughout the nation. Similarly, they provide the basis for intrastate aids and for special aids to central cities. Harvey Brazer poses the problem of the central cities in these terms: "To the extent that suburban communities, through zoning regulations and discriminatory practices in rentals and real estate transactions, contribute directly to the concentration in the central city of socio-economic groups which impose heavy demands upon local government services, they are, in fact, exploiting the central city."[11] Aids to central cities to meet these problems are increasingly being urged.

Intercommunity flows aside, under the present categorical aid programs the local policy official may be choosing among programs in which the local tax commitments vary widely per dollar of program expenditure. At one extreme, the grant may approach 100 per cent of costs, at the other there may be no outside aid. The local policy maker is weighing the unevenly quantified costs against benefits. Grant-induced transfers of resources may result, with more of the public service offered where the grant funds are most favorable and least for those services that are not aided. Tables 2, 7, and 8 indicate some of the variations in state aids within a metropolitan area. To meet this problem, unconditional fiscal or "block" grants by the state to localities are proposed. "In contrast to the grant-in-aid, or the shared tax, the block grant provides complete budgetary freedom at the local level, since the use of the funds is not restricted, and ensures stability."[12]

Table 7. Matching shares: selected federal grants

Program	Non-federal share (percentage)
Highways	
Interstate and defense	10
Primary, secondary and urban	50
Civil defense	
Evacuation and shelter planning	0
Vocational education	50
School construction, federal impact areas	0
School maintenance, federal impact areas	0
School guidance and counseling	50
Science, mathematics, and foreign language laboratories	50
Public health grants	50
Water pollution control	Variable 33-1/3 - 66-2/3
Hospital and medical facilities	Variable 33-1/3 - 66-2/3
Child welfare service	0
Vocational rehabilitation, special projects	33-1/3
Urban renewal projects	33-1/3
Airport construction	50
Urban relocation	0

[11]Harvey E. Brazer, "Some Fiscal Implications of Metropolitanism," in Metropolitan Issues: Social, Governmental, Fiscal, Guthrie S. Burkhead, ed. (Syracuse University, Maxwell Graduate School of Citizenship and Public Affairs, February, 1962), p. 77.

[12]Governor's Minnesota Tax Study Committee, Report of the Committee (Minneapolis: Colwell Press, 1956), p. 527.

Table 8. Local matching shares, an illustration: New York State

Program	Local shares (percentage)
Public health work	50
Laboratory services	50
Adult poliomyelitis	50
Tuberculosis care and treatment	50 up to $5 per day; 100% of excess
Medical care for rehabilitation of physically handicapped children	50
Mental health	50 up to $1.20 per capita; 100% of excess
Probation services	50
Highway aids (generally)	none
Town highway improvements	variable down to 25% min.
Snow removal	0
Public works	
Plans for local improvement	50
Urban planning assistance	50 of excess over federal aid
Youth bureaus and recreation	50 within limit
Veterans' services	50 within limit
Housing and urban renewal projects	
Low rent housing	50 in excess of federal aid; payable in kind
Capital grant	50 in excess of federal aid; payable in kind
Open spaces for recreation	
Land acquisition	25
Costs of transportation of physically and mentally handicapped children	50
Aid for services for emotionally disturbed children	50
Vocational education and extension boards	50 (100% in excess of limit)
Community colleges and 4-yr. colleges	
Operating	33-1/3
Capital	50
Aid for experimental educational programs	variable up to 50
Aid for education for pupils from culturally deprived groups	50
Use of educational TV	
Construction	50
Operation	escalator clause 50 1st year 90 5th year
Recreation program for elderly	50 within limits
Care of juvenile delinquents	50
Welfare	
Special categories	50 of excess over federal aid
Home relief	50 down to 20% on bracket of costs

The relevant factor, however, is not the matching share on the average but the matching of the incremental expenditures. Under present grant procedures and provisions, matching on incremental expenditures may differ substantially from average matching. Under most state grants aid to a local government does not increase as local expenditures for a program are enlarged. Grants distributed among local governments on such bases as population (total, school age population, etc.), or land area, or a uniform amount for each jurisdiction, do not result in more funds to a community, if the community expands its aided public service. Additional expenditures (in the first instance at least) must be financed out of local taxes. The decision to spend more requires a concomitant decision to raise more local taxes.

Similarly, grants distributed among localities on the basis of the number of pupils enrolled in public schools, cases treated for tuberculosis, or miles of highway do not necessarily mean more state funds if local expenditures are increased. Two communities with different expenditures per pupil enrolled would receive the same amount of state aid. Only if the level of public service increases the number of pupils, or number of cases treated, is state aid enlarged, and then only by a redistribution of the state appropriation among the aided communities. Equalization aid works in the same way once the minimum local tax effort requirement is met. Additional local expenditures are financed by additions to local taxes, rather than through more state aid.

The only grants that result in an increase in aid proportionate to increases in local expenditures, are those that call for state or federal matching of local expenditures in some fixed ratio. Federal grants for slum clearance, public housing, and waste treatment works are of this type. While the state public assistance grants also generally are shared proportionately, statewide standards limit local decisions on program expansion.

We need to assess simultaneously the two problems: (a) the possible misallocation of resources resulting from limited use made of differential grant offerings for specific functional expenditures, and (b) under-allocation of resources when benefits accrue to outsiders (or over-allocation when services are financed by outsiders). An unconditional grant to a city, which reinforces decisions of local voters, will not assure adequate services for the commuting nonvoter. It may be that a categorical aid in an amount based on the value of outsiders' benefits from a particular service would improve the allocation of resources. It may turn out on investigation that one of the main criticisms of the present categorical aids as currently designed is not that they have unequal matching shares but that the aids for specific functions are not sufficiently different to offset the variations in spillover of benefits.

To leave the statement here would be to greatly understate its complexity. Jesse Burkhead in another context noted: "We do not have a regional social diseconomies board with authority to parcel out the spillover gains and losses to effect a rational allocation of resources in accordance with improvements in welfare among subunits of the region."[13] A categorical grant system related to these spillovers would require that a state diseconomies board be set up as well, with authority to give differential aids to local jurisdictions.

Moreover, functional programs are closely interrelated. Expenditures in one function affect expenditures on others in a variety of ways. Public services in one functional area:

---tend to result in a saving in another functional area (e.g., improved educational services reduce juvenile delinquency; improved mass transportation reduces costs of traffic congestion; improved public housing reduces costs of a public TB hospital and of other treatment).

---buttress and make more effective services in another area (e.g., multiple purposes of water resource programs; health services, school lunch program, and aid-to-dependent-children payments improve effectiveness of public schooling; old age retirement payments back-stop housing programs for elderly).

---

[13] Jesse Burkhead, Comment in Design of Regional Accounts, op. cit. (footnote 8), p. 68.

---require the expansion of outlays in other functional areas (e.g., new streets and highways open up new areas for development and require expansion of water and sewer mains, schools, etc.).

---make services in another area inefficient or obsolete (e.g., inadequate staffing of a juvenile court increases length of stay at detention villages. A misplaced highway may mean tearing down a school building or eliminating a public park, and can destroy plans for the use of land, as well as increase congestion).

---uncover new program needs in other functional areas (e.g., slum clearance leads to relocation services).

These relationships also have their negative counterparts. For example, lack of public services in a functional area may result in additional outlays in another functional area.

While the programs are closely interrelated, each has its own definitions of a beneficiary group, its own clients. Welfare programs provide aid on the basis of individual family resources in relation to the standard of need. Housing programs have income tests. Medical care is provided on the basis of "medical indigency," public education uses a test of "residency," and the school lunch program, at least in some places, provides free lunch for those living in specified dwellings or neighborhoods. Some public health services are available to all comers, others to those "unable to pay."

Thus, many functional agencies, each with its own primary "mission," are providing the urban family with a bundle of closely interrelated public services. Public programs to meet the requirements of youth, of the aged, of low income families, of the suburban commuter, or of new migrants in cities call for the combination of a range of functional services.

To take one illustration, the large in-migration of population into metropolitan areas and the concentration of nonwhite populations in central cities create many of the difficulties currently facing those concerned with the effective operation of functional programs. As the Advisory Commission on Intergovernmental Relations states: "Population is tending to be increasingly distributed within metropolitan areas along economic and racial lines. Unless present trends are altered, the central cities may become increasingly the place of residence of new arrivals in the metropolitan areas, of nonwhites, lower-income workers, younger couples, and the elderly."[14] In the period between 1950 and 1960 the nonwhite population in Baltimore rose 45.3 per cent, in Chicago, 64.4 per cent, at the same time that total central city population declined.

Groups concerned with poor health and costs of public medical care, with slums and relocation, with low educational attainments, and with losses due to drop-outs from school, with crime and delinquency, with mounting welfare costs, are turning to function-by-function solutions to urban problems created by in-migration of families from non-urban settings. Extension of the public assistance grant is urged to "provide incentives to recipients of public assistance to improve their conditions" to become self-supporting. The manpower and training act of 1962 is proposed to

---

[14]Advisory Commission on Intergovernmental Relations, op. cit. (footnote 5).

give individuals an opportunity to acquire new skills in order to alleviate the hardships of unemployment. New federal-state aids are proposed to wipe out adult illiteracy and steps are being taken to improve the opportunities for education of children from "culturally underprivileged home environments." Federal housing programs are being extended to improve the quality of living of the families housed. And the concept of planning is being extended from physical facilities alone to planning for use of facilities and services by the urban family.

Functional fragmentation of solutions to public policy problems, whether it be to help adjust newly arrived families to urban living or to meet other problems, creates an urgent need for co-ordination among governmental programs. These problems, however, are not resolved simply by a block grant of funds; co-ordination among functional agencies still is the primary requisite. When categorical grants are offered, surely the types of categories need to be carefully reassessed to determine whether they are appropriate. But more importantly, ways must be found to assure the effective combination of services.

One possible method of co-ordination is to condition each of the categorical grants on the formulation of a community-wide plan of operation, and require approval of this plan by each of the granting agencies.

The carrying out of such a plan would be strengthened by a complementary provision calling for the horizontal co-operation of vertical functional agencies--co-operation on a specific problem of concern to many agencies. This could be achieved by the designation by the mayor (or other appropriate governmental official) in community of a "single directing agency," with authority to work with other local departments and to co-ordinate those activities of the several departments relating to a single problem, for example, adjustment of families to metropolitan living.

As an additional step toward a unified policy, transfer of funds among programs might be authorized. By these or substitute methods co-ordination could be achieved and costly duplication of efforts could be avoided. A concerted effort along the lines summarized above would achieve over a period gains in the quality of the population in the cities and could contribute to the economic growth of the cities, an expanded tax base, and a reduced load of public dependency.

4

General and Specific Financing of Urban Services

by William W. Vickrey

The purpose of this paper is to re-examine the degree to which municipal services are financed from (1) general purpose taxes, levied without close attention to the way in which the taxpayers benefit from public services or affect the costs of rendering them, and (2) specific taxes, fees, and prices that attempt to reflect these costs and effects more closely. To some extent the concern with bringing payments more closely into line with costs and benefits is related to concepts of equity, in that it is conceived to be in some sense proper that those who enjoy benefits or give rise to costs should, in the absence of countervailing considerations, pay accordingly. But more weight, on the whole, is given in the current investigation to the possibility that such correlation of charges with costs and benefits can be made to increase the efficiency with which services are utilized, prevent waste, and in general improve the patterns along which our mushrooming metropolitan areas will grow. Indeed, it is this latter consideration that leads to a dissatisfaction with a mere statistical or average balance or proportionality between benefits and contributions and an insistence on a greater precision of detail: situations can easily arise in which groups of more or less similarly situated individuals share a cost equally and hence no individual is in fact treated unfairly, yet if the institutions are such that no one person can reduce his share in the cost by suitably economizing or restraining himself, the amount of the service demanded and supplied may be grossly excessive.

GENERAL CRITERIA

In determining whether an attempt should be made to pay for a municipal service by means of a specific charge, a number of general principles or criteria can be referred to; their specific impact varies, of course, from case to case. One of these concerns the distributional impact of the charge relative to that of the general tax that it might displace. In many cases this differential will be small enough or uncertain enough in its direction to be considered a wholly secondary matter, but in particular cases it is important and even paramount. In New York, for example, a straight increase in transit fares can be considered almost tantamount to a poll tax in its incidence, and certainly far more regressive in its distributional impact than even the sales tax. The distributional impact of an attempt to finance educational, hospital, or welfare services entirely by direct charges would obviously be so unacceptable as to preclude such a solution, at least in its simple and direct form.

Another general principle that can be appealed to is the extent to which the proposed charge can be related to benefits derived from the service in a way which will appeal to concepts of equity. This immediately raises the question of how to measure the benefit: should it be in terms of the cost of providing the service to the individual concerned or in proportion to the amount that the individual would pay rather than go without the service altogether? To what extent, for example, should water charges be based on the income of the user, or be differentiated according to the cost of his obtaining water from another source? It seems fairly clear that answers given to such questions will vary widely from case to case and that the variation will have to be explained on grounds other than a general adherence either to cost or to utility as a basis.

Indeed, one factor that will often enter into the general attitude taken is the nexus, as perceived by the public, between the payment made and the benefit received. In many cases this perceived nexus will be significantly influenced by historical development. A new service accompanied by a new charge is likely to generate a vivid conception of a quid pro quo; a new charge made for a pre-existing service is more likely to create a feeling of inequity, even if the charge is a necessary means of preserving the value of the service, as in the case of the ferry that is worthless as long as it is free (because then queues accumulate until it is just as quick or convenient to detour via a bridge or tunnel), or where transit services are subsidized in part from funds obtained from highway use charges. Another element in the perception of this nexus is the manner of payment: a toll is immediately visible as an out-of-pocket cost financing a distinctive service; a gasoline tax is slightly less so; payment for a sewer connection may evoke different attitudes, even though computed on the same base, according to whether it is merely included in a global tax bill to a municipality performing all of the various functions itself, or paid as a separately itemized item on a consolidated tax bill, or paid separately as an entirely distinct transaction. All of these may have a significant effect on the way in which the relation between benefit and burden is perceived by the general public and in the degree of acceptance of this relationship.

More attention is to be paid here, however, to the matter of allocational efficiency. This in effect means extending the concept of marginal cost pricing as far as possible into the realm of municipal services with the intent of attaching, to each choice by individuals that affects municipal operations or that has an impact on others in the community, a differential charge that will properly represent these impacts to the individual. Ideally, this charge would serve to co-ordinate decentralized decision making on the part of individuals into a harmonious whole. In many areas, however, it will be an adjunct, though an important one, to more centralized planning activities as represented by building codes and zoning ordinances.

Moreover, even in the absence of pressures arising from considerations of progressivity and of equity perceptions, it will in practice be impossible to approach even very closely to this theoretical ideal. The relative social costs occasioned by alternative choices contemplated by individuals cannot be measured, in many cases, with anything like complete accuracy, and in other cases the costs of carrying out the measurements would be such as to outweigh the benefits. Even where the costs can be ascertained with adequate accuracy, the costs of assessing the corresponding charges may be great, and in some cases the terms of the assessment might tend to become so complicated as to be beyond the comprehension of the individual making the decision, and thus be ineffective in securing the improvement in the allocation of resources that was the original *raison d'être* of the more elaborate assessment. And finally the fact that in nearly all cases, even with the fullest possible utilization of specific charges, municipalities are almost always in such pressing need of funds that they must have recourse to general taxes that themselves have adverse effects on allocation. Consequently, by virtue of the general principle of "second best," if any specific charges are to be made, they should in nearly all cases be designed in part to contribute to the public treasury over and above the amount that would flow in on the basis of charges strictly reflecting marginal costs.

SPECIFIC CASES

General principles are seldom as enlightening in the abstract as in their application to specific cases, and accordingly some specific examples are discussed in the remainder of the paper which will serve to illustrate their application and explain their meaning. In considering specific cases, we will include on the one hand some items that may not ordinarily be thought of as municipal services, but which would lend themselves to financing by methods ordinarily associated with municipal finance; and on the other hand some municipal services, including some of the more important ones, will not be considered because of the inherent difficulty of applying to them the methods being considered here.

Fire Protection

Fire protection accounts for some 7.6 per cent of all general expenditures by cities. It serves here as a striking example of the difference between benefit and cost as a basis for charging.

Actually, fire protection is considered most appropriately paid for on the basis of property assessment, and in terms of benefit this is a fairly good basis. Benefit can be roughly measured in terms of the reduction in insurance rates on protected as against unprotected property, and in terms of the enhancement of land values in view of the provision of the protection to any structures built on the land. The value of land plus improvements would accordingly seem to be a good measure of benefit. Even here, of course, there are cases where increased property value fails to indicate increased benefit: replacing an old, obsolescent building with a modern fireproof one may lessen the amount of benefit obtained from fire protection even though the value of the property be increased.

Viewed from the cost side, however, the picture is entirely different. Providing a given grade of fire protection to an area is almost entirely a matter of providing an engine company within a suitable distance of the property, or more operationally, within a suitable number of minutes of travel time. The National Board of Fire Underwriters, in setting standards for fire protection, requires as a rough rule of thumb that in residential areas a company be stationed within about 1.5 miles for a property to be considered adequately protected; for industrial and commercial property, apartments and the like, the distance is shortened to 0.75 mile, while for areas containing only widely scattered residences a distance up to 3 miles is tolerated. There are in addition minor differences in the cost per engine company: in business and industrial areas a complement of seven men on duty is considered normal, whereas in residential areas five men on duty may be considered sufficient; in addition the property occupied by the fire house may be more valuable in the former case. Roughly speaking, therefore, the ratio of the cost of protecting an acre of residential area and that of protecting an acre of business area may be about one to five (leaving the fringe areas out of consideration, for the moment). This is a much smaller range than the corresponding range in assessed values, in most cases.

From the point of view of resource allocation, however, it would not be sufficient to remain content with property values as the basis for paying for fire protection even if it could be shown that these costs in fact varied in proportion to assessed value. An increase in the value of the improvements in a given area, through new construction or otherwise, does not in any significant way increase the cost of furnishing a given grade of protection to the property already there, even though it may make it worth while to provide a higher and more costly grade of protection. Even though in principle the construction of new buildings in a given area may increase the frequency of fires within the area and thus might increase the probability that the equipment might be out on another call when fire breaks out in a given property, this is normally a very minor factor, made quite negligible in most cases where alternate protection is normally available from adjacent fire companies. The basic act that is causally related to the need for added fire protection is the occupancy of land in the protected area. It is the exclusive occupancy and not the nature of the occupancy that counts: if an acre of land is used for tennis courts or a clay products depot, involving of itself almost no fire hazard, the displacement of the occupancies that might have used this land to outlying areas where new fire protection will have to be provided is a cost that should properly be assessed against such uses. Similarly it makes no difference to the cost to be assessed against a given acreage whether it is developed with residences on half-acre lots or row houses cheek-by-jowl, even though this might make a difference to the way an insurance company might rate the hazard.

Accordingly, the appropriate way to charge for fire protection would be on the basis of area. Possibly some gradation in the charge might be made in terms of distance from the fire house, but in most cases this would be a negligible factor. An exception might be made where the protected area dependent on a given fire house includes both residential and business property: areas zoned for business and located within the .75 mile distance might be assessed at a higher rate based on the higher degree of protection offered and the need of business for this higher grade of protection if it is forced to locate elsewhere by inferior occupancy of the business-zoned area. In this case it would be the nature of the zoning rather than the nature of the occupancy that would be the appropriate basis for the distinction in the charge. Moreover there is in a sense a joint product problem here: a given fire house required to provide residential-grade protection for the 1.5 mile radius necessarily provides incidentally for business-grade protection within its .75 mile radius. On the one hand if the inner zone thus generated is greater than the demand for business sites, the business-grade protection is a by-product to be provided at no marginal cost, the residential occupancy bearing the full burden; on the other hand if the business demand increases so as to require the provision of an additional fire house, the additional residential protection, if enclosing some unoccupied but developable land, is likewise a by-product, and the full cost of such a fire house should be borne by the business protection area.

Translating the charges thus indicated into an actual tax is of course another matter. In the long run it would probably not make too much difference if the assessment were made on the basis of land value rather than land area, as any difference between the two forms of charge could readily be capitalized into the land value. In the short run transitional effects would arise, but consideration of these effects will have to be relegated to another analysis. In any event it is clear that on a cost basis, the basis for assessment is clearly land value, exclusive of improvements, in spite of the fact that at first glance the benefit basis would seem to indicate that it is improvements that should be assessed.

Another factor associated with fire could be the basis for the assessment of charges--the external economies associated with what are technically termed "exposure fires," i.e., fires not starting on the premises under the same ownership. Individuals who take precautions against fire are not only reducing their own risk, but that of their neighbors. To the extent that they are reducing their own risk in recognizable ways, they may qualify for lower insurance premiums, but since neighboring properties are insured by independent companies, and indemnities are paid regardless of the source of the fire, these rating allowances of necessity include no allowance for reductions in the likelihood of contagion or exposure losses. It would accordingly be of some merit for property tax assessments to make some small concession to structures that are fireproof or are sprinklered or are otherwise protected, over and above what building codes may make mandatory. The tax concession could even be more widely applied than building codes, since the latter are often full of grandfather exceptions; in some cases the tax concession might even motivate an improvement where otherwise the owner would be willing to hide behind an excepting clause.

Too much should not be made of this, however: over the period 1953-59 the property loss claims from exposure fires amounted to $221 millions, which is 7.2 per cent of losses from known causes, which in turn was 45 per cent of total losses. Since exposure fires probably constituted a relatively low percentage of the "causes unknown" category, such losses probably amounted to not more than 4 per cent of the total. Though the principle may be applicable, the practical consequences are _de minimis_.

Water Supply

In water supply we are faced with a wide range of situations as to source of the bulk supply, but the over-all characteristics of distribution tend to be somewhat more uniform, and we will discuss the latter first.

In some fortunate areas, reasonably pure water may be available from a nearby source in ample quantities, so that the main problem is one of distributing this supply to users. Moreover the economies of scale in water main size are so substantial that it is only a minor oversimplification to say that the cost of the mains is proportional to their length and relatively independent of the required volume of flow. Another factor tending to warrant such a simplification is that for fire protection purposes a certain minimum size main is required to provide the flows needed for fire-fighting purposes.

The problem is how to relate the total length of main required to the nature of the occupancy of the area. It is tempting to say again that area is the critical factor, but this is overruled by the observation that if we take two communities that are laid out in a similar pattern, but on different scales so that streets and mains are twice as far apart in the one community as the other, while lots are four times as large, the second community will only require twice the length of mains. The simplest rule would be to use the front foot as the unit; this however produces awkward results when applied to corner lots, but even so is perhaps the best simple rule available. One can of course expect any minor variations in the system of charges to become capitalized in the price of the lots, in this case, so that the corner problem is perhaps not a crucial one. The same is true of variations in the relative amount of frontage associated with various lots due to odd shapes, curves in the street, and the like. It would be tempting to try some more general linear measure of lot size, such as the square root of the area, or the maximum diameter, but these would produce results that make the aggregate charge for a group of lots vary according to the way the total area is subdivided in ways that bear no very close relation with the length of mains required.

In smaller communities endowed with an ample source of supply this may be all that marginal cost pricing requires: there would remain, in addition to the cost of the mains, the cost of the collection and purification system and the transmission aqueduct from the source to the edge of the community. In many cases the incremental cost of added capacity may be so low as not to warrant the cost of installing and reading water meters. These intra-marginal costs not covered by explicit charges for mains can then be covered out of general revenues, or from any other source on the basis of "least harm." One source would be surcharges on the cost of the mains; not only would this probably not be too much of a distorting influence, but it would have considerable public appeal on equity grounds. Actually, data for 1959 indicate that over-all expenditures for water supply amount to $1,423 million as against revenues of $1,201 million, indicating some support from general revenues, though since expenditures are largely on a cash rather than an accrual basis, with considerable confusion between current and capital outlays and probably a considerable understatement of real interest and amortization charges, this conclusion should be made with caution.

In many cases, however, especially in the more densely populated areas and arid regions, the incremental cost of the gross supply will be a significant factor. The problem is complicated by the large size of the lumps in which increments to supply are often available, the great durability of the facilities, and by the significant seasonal and random fluctuations in the supply furnished by given facilities. Under such circumstances charges according to the amount of water used are obviously in order, but have in the past been justified far more on grounds of equity than on grounds of controlling the use. Indeed, the idea of using water charges to ration use seems quite at variance with typical attitudes of water supply engineers, who seem to view their function as one of providing an ample supply almost regardless of the cost of meeting whatever standard they decide to set, and limiting the charges to the user to whatever minimal rates prove necessary to finance the scheme over all, regardless of what the incremental cost of the particular supply involved happens to be.

Reluctance to use water charges as a means of rationing use is illustrated in the fact that rates generally remain much the same from one season to another and from one year to another. To some extent this is the result of sheer administrative inertia, but to a considerable extent it represents the view that to raise rates at a time of shortage represents exploitative profiteering that is considered undesirable and even immoral, regardless of whether the beneficiary is a private individual or a public body. Nevertheless, to an economist, at least, the possible improvement in the allocation of resources through such variation in rates should be fairly obvious, the main question being how difficult such variation would be and how important the benefits that would result.

Variations in supply may arise either as a regular seasonal phenomenon, or irregularly as a result of variations in rainfall or large accretions of capacity. In all three cases variation in rates would improve resource allocation, but the difficulty of applying the changes varies considerably. One difficulty common to all three situations is that meters are usually read quarterly, and often on a rotating schedule, so that it would be difficult to bring the impact of the rate change to bear on all consumers simultaneously. In a system having large reserves, where a situation of shortage or abundance can be expected to persist over several quarters, or in the case of the addition of large increments of supply, it may be sufficient to continue a quarterly pattern, simply prorating the consumption indicated for a given quarter into the portions of the quarter falling into the low and the high rate periods. This works fairly well with utility bills at present, where rate changes occur, but in such cases the rate changes are relatively minor and the principal issue is one of equity; it would be relatively less satisfactory where rate changes are of a magnitude intended to be sufficient to affect consumption. Then the discriminations would be somewhat more severe and the dilution of the incentive to economize during the latter scarcity part of a quarter, by reason of the fact that such consumption would be prorated back to the earlier or abundance part, (and vice versa in case of rate reductions) would be of significant concern. Ideally, extra meter readings should be arranged as nearly as possible to the time of the rate change, so as to minimize this prorationing effect; possibly this might be done economically by combining them with electricity and gas meter readings. Self-reporting by the consumer, such as is often practiced either for interim readings or where the regular reader was not able to gain access to the premises, would seem somewhat less feasible here than where no rate change is taking place: with constant rates there is ordinarily no significant incentive for the consumer to falsify or fudge his report of the meter reading, and even where there is, the consumer would often have to be fairly sophisticated and prescient about his future consumption to take advantage of the opportunity; where a substantial rate change is involved, both the magnitude of the incentive and the clarity to the consumer of the direction in which it is to his advantage to misreport would operate to make misreporting more of a problem.

One objection to fluctuating rates is that they make it more difficult for the consumer to budget, particularly with rates that fluctuate with unpredictable variations in rainfall. The problem would be less significant with preditable seasonal fluctuations in rates. Unlike a private utility, a municipality is in a position to offset changes in water charges with changes in the general property tax rate, so that what the average consumer would gain on a given occasion in lower water rates he would lose in higher property taxes, and vice versa; the incentive for conservation of water in times of shortage would remain. In effect, a substitution effect is generated while the income effect is minimized.

With water more than with most other utility services, the tenant who turns the tap is often insulated from the impact of water rates by their inclusion in his over-all rent, and for this reason rate variations may be less effective here than with other services. Such use is however relatively inelastic to price in any case, and control at such points is hardly worth while. There remains, in any case, the incentive for the landlord to pay special attention to leaks and other outright wastage and for economizing in large-scale industrial use, lawn watering, and the like.

One need not necessarily maintain that rate changes can entirely take the place of water conservation campaigns in times of threatened shortage, but high water charges should provide a significant reinforcement for such campaigns, if they become necessary. Certainly such charges provide a more efficient means of rationing than such methods as the prohibition of certain uses at certain times, or in extreme cases the shutting off of the supply in various areas at various times, with the attendant danger of contamination and increased fire hazard.

Variable rates should permit considerable economies to be made in scheduling expansion of supply. Given a means of efficiently controlling the demand in cases of drought of rare intensity, it will be less necessary to expand supply quite so far to take care of such contingencies. Similarly, it will be less necessary to plan new additions to supply quite so far in advance, since a faster than anticipated growth in demand or shortage due to subnormal rainfall can be adapted to more effectively. On the other hand, new additions to supply can be used more fully and more promptly upon their completion if charges are reduced or eliminated as long as the supply is ample.

However, a policy of raising charges prior to the completion of a new project and lowering them when it is brought into production means a drastic change in traditional modes of financing public works. To the extent that funds are accumulated out of the earlier high rationing charges, the project will be more nearly on a "pay-as-you-go" basis; on the other hand, there will be no funds flowing from water charges to pay off the balance of the cost until such time as the new addition is being fully utilized and rationing charges are again needed. Whether or not this pattern can be followed in its pure form will of course depend on the over-all fiscal and financial pattern of the municipality, but substantial changes in financial procedures are indicated in any case, even if the limiting situation is not a feasible one.

There is one further difficulty with drastic variation in rates over fairly long periods: consumers may be induced to install water-using equipment, such as non-recirculating cooling equipment, on the basis of current low rates only to find that when rates are later raised their investment turns out to have been unprofitable. In principle, of course, consumers should be put adequately on notice of the likelihood of water rates being increased in the future. Fortunately, most decisions of this sort that involve substantial fixed capital and that would be significantly affected by water rates are of a commercial or industrial nature. In such instances the decision-maker can be presumed to be sufficiently sophisticated to take account of such a prospect.

Transportation Facilities

The over-all picture of the finances of urban transportation conceals a great deal of inefficiency and distortion within an aggregate picture that seems spuriously close to being in balance. Aside from a few outstanding instances of substantial subsidy, such as the New York subways and rail commuter service, it appears superficially that the transit rider is about paying the costs of the service he uses. Likewise, for motor traffic as a whole, revenues from user charges seem to be roughly in balance with outlays on facilities: in 1957 total tax revenue from all motor vehicle related sources, including licenses, fuel taxes, manufacturer's excises, tolls, and parking meter revenues amounted to $8,162 million; expenditures directly on highways amounted to $7,931 million, not counting that part of local police expenditures of $1,290 million that could be regarded as spent for traffic control (roughly one-tenth, judging from the number of police assigned to traffic duty in New York City), or the portion of the $562 million of "other sanitation" expenditures that could be regarded as spent for snow removal and other traffic-related purposes.[1] On this basis one would be inclined to doubt whether there is any large scale misallocation of traffic between various modes, or substantial under-or-over-utilization of facilities.

When looked at in more detail, however, the picture is quite different, especially when full economic costs not represented adequately in the financial accounts are added in. If we attempt to separate out the urban component of the motor vehicle revenues and expenditures, we find that in 1960 the forty-three largest cities spent $600 million on highways, exclusive of police and sanitation expenditures for traffic related purposes. Of this only $289 million was financed from vehicle user charges in terms of any explicit flow of funds: $155 million from state highway grants and $134 million from city licenses, parking fees, and user charges, leaving $311 million, or 52 per cent of the total outlay to be defrayed out of general revenues.[2] In 1957, all cities spent $2,941 million on streets and highways, of which only $1,471 million, or just 50 per cent, was derived specifically from use-related charges.

To get closer to the allocation of resources, however, we cannot stop here for the relevant matter is not through what channels the funds flow, but whether, on balance, the charges the motorist pays, to whatever governmental agency, correspond to the costs they occasion, in whatever form and by whatever agency these costs are borne. (Indeed, one could argue that the general purpose grants from state to local governments include some highway funds.) Figures of this sort are harder to come by. However,

---

[1] U.S. Bureau of the Census, Census of Governments, 1957 (federal excise taxes added).

[2] U.S. Bureau of the Census, Compendium of City Government Finances in 1960.

in a recent study of the Philadelphia metropolitan area, it was found that in 1957 total motor vehicle tax charges allocable to use of motor vehicles in the city of Philadelphia amounted to $30 million to all levels of government, but that motor vehicle expenditure by all levels of government for facilities within the city amounted to $46 million. Thirty-five million dollars of this was spent by the city itself, of which only $6.5 million was financed by grants from federal and state highway funds, and by user charges (chiefly parking meters), leaving $28.5 million to be financed from general revenue sources.[3] While this is only one instance, there is no obvious reason to suppose that the total cost of facilities being provided in Philadelphia is significantly higher relative to use than in other metropolitan areas, nor that the level of charges being paid by motorists is significantly lower. On this basis motorists in large cities appear to be paying only about two-thirds of the current expenditures made for the facilities they use.

To obtain the true economic cost of urban traffic facilities, however, it is necessary to go even further than this and substitute, in place of current capital outlays, an appropriate rental charge representing interest and depreciation or amortization on the value of the facilities used. The current capital outlays to be substituted for, as representing provision for future use rather than for current use, amounted to $351 million out of a total of $600 million for the forty-three largest cities; for all cities in 1957 the figure is $754 million out of $1,753 million. But what to put in place of these figures is something of a problem. Farrell and Paterick, of the Bureau of Public Roads, put the value of the depreciated investment in urban street and highway improvements at $10.2 billion as of January 1, 1953; $11 billion would seem a reasonably rough figure for 1957.[4] A figure of $8.25 billion for the value of land in city streets was produced by the Federal Trade Commission for 1922;[5] a more recent estimate appears not to be available but by considering trends in property values, a 1957 value on this basis of $21 billion seems roughly reasonable. At 4 per cent interest on the total value of $32 billion, plus 2 per cent amortization on the depreciated improvements, this gives $1,500 million to be substituted for the $754 million of current capital outlays in figuring true economic cost. The full annual economic cost of city street use thus comes to $2,319 million, as compared with actual outlays of $1,573 million, exclusive in both cases of police and sanitation. Even at this, the computation includes no equivalent for property taxes or corporation income taxes that would be covered by rental payments for the use of privately constructed property requiring comparable amounts of land and construction. If property taxes were figured at 1 per cent on the full value (in 1956, assessments of $280 billion were estimated to represent 30 per cent of full value, or about $930 billion; property tax collections were $11.7 billion)[6] $326 million would be added to the bill; in addition if as little as 20 per cent of the investment were financed by equities, with a net return to equity holders of only 4 per cent, the corresponding corporation income tax of 52 per cent would add $277 million to the cost. With these elements added in, urban vehicular traffic is found to be paying $1,050 million in charges as compared to

---

[3]Philadelphia Bureau of Municipal Research, Improved Transportation for Southeastern Pennsylvania (May, 1960), pp. 294, 314.

[4]Proceedings of the 32nd Annual Meeting of the Highway Research Board, January 1953.

[5]National Bureau of Economic Research, Studies in Income and Wealth, Vol. 12, p. 547.

[6]U.S. Bureau of the Census, Statistical Abstract of the United States, 1961, pp. 403, 418, 419.

roughly $2,916 million that users of comparable resources in the private sector of the economy have to pay.

Obviously, if there is a way in which the cities can collect an additional $1,866 million from urban vehicular traffic so as to bring the charges more nearly in line with the costs, it would not only help to improve the resources allocation pattern but would ease the financial pressure on municipal governments very substantially. (Total city revenues from all sources were $13,748 million in 1959.) Unfortunately, there has hitherto been no easy way to approximately triple the charges payable by the urban motorist without excessive interference with the smooth flow of that part of the total traffic that has a legitimate and urgent need for the use of the city streets. It appears, however, that with the aid of modern electronic equipment it is now feasible to equip each car using the urban streets with a cheap and rugged response block by virtue of which a record can be made each time such a car crosses a zone boundary, without interfering with the flow of traffic, and these records can be processed on electronic computers to produce a bill which can be made to vary appropriately with the costs occasioned by the indicated movements of the car. It remains to be seen, however, whether what is economically sensible can be made politically palatable.

Some demurrer to the assignment of all city street costs to vehicular traffic is often entered on the ground that in addition to carrying through traffic such streets also serve as "access" to the adjacent property, and that the cost of such an access function is properly chargeable against the property owners rather than against the users as such. To be sure, if conditions are such that the roadway is no more elaborate than that which would be required to provide a mere access, and if traffic conditions are such that there is in fact no interference with other users of the roadway during such "access," then the marginal cost of such use is effectively zero in the short run, and charging the entire cost of the access street against the property owner would be conducive to unrestrained use of the uncongested facility, so that efficiency would be served. The bulk of the cost, however, particularly in view of such factors as the use-related character of much of the outlay for renewal of pavement and the high proportion of the property values accounted for by the downtown areas, is more nearly chargeable against users than abutting property owners. Even when a vehicle is performing an access function, its impact on traffic and on costs may be substantial; the amount of these costs is a function of the movement of the vehicles performing the access function and is not related in any direct way to the value of the property accessed. While some allowance might legitimately be made for the access function of low-traffic residential streets, this allowance would be small. Moreover, even on equity grounds, one could well raise a question as to whether the charge for the provision of common access facilities would not be more fairly allocated according to amount of use, rather than according to the value of the property accessed.

For example, even where a cul-de-sac is used almost solely for access purposes, with no traffic movement other than that destined to one of the abutting property owners, if conditions of use are such that there is interference between users, some means of bringing the cost of this interference home to them directly is needed. This can be done conveniently if the costs associated with providing the area and the pavement are assessed against users in proportion to use. It is both inequitable and inefficient to charge these costs to the abutting property owners under these circumstances, making the abutting occupant whose business requires relatively little use of the access facility pay just as much as his neighbor whose operations seriously interfere with the ease of access of the adjacent occupants. Some of the most serious congestion in New York is the result of the way in which "access" is had to firms in the garment district.

The special problem of the rush hour. The disparities in treatment between various forms of transportation are even further aggravated by the nature of rush-hour conditions. Nearly all forms of rush-hour transportation are grossly underpriced. In 1951, at a time when the fare of 10 cents on the New York subways was just about covering operating expenses, it was estimated that the marginal cost of a moderately long rush-hour ride was somewhere between 20 and 40 cents, possibly even higher in some cases. The differential between rush hour and non-rush hour or between rush-hour costs and average over-all costs varies widely as among alternative forms of service, so that even if various forms of transportation were on an equal footing over the entire week, either each paying its way or being subsidized by about the same proportion, this still would not bring about an economical allocation of traffic among the various modes during the peak hours. For one thing transit systems can handle overloads in general much more readily than highways, by putting more persons into the same vehicle, and while this causes some deterioration in the service, sometimes to a truly inhuman degree, there is no complete breakdown. On the highways, on the other hand, there is no mechanism whereby increased traffic flow results in more persons riding per car, and there is a rather strong tendency for the system to jam up once a critical flow has been reached. Another and more important factor is the distribution of traffic through the day. At one extreme, commuter railroads handle half of their total traffic during the fifteen peak hours per week; for subway traffic the figure is about one-third, while for expressway traffic on routes to anu from the central city the proportion is about 18 per cent. This means that when costs are averaged over the entire traffic flow, the rush-hour traveler, who should bear the bulk if not the entire amount of the cost or providing the capital facilities, is able to shift over four-fifths of this burden to non-rush riders if he is an auto rider, but only two-thirds if he is a subway rider and only half if he is a rail commuter. Differentiation of the charge between peak and off-peak use is thus not only a matter of encouraging peak use and discouraging off-peak use, but it is also a matter of inducing the selection of a suitable mode of transportation given the time and volume of travel.

Accordingly, there can be no efficient solution to the urban traffic problem that does not include provision for charges on automobile use that are differentiated according to time of day. The most straightforward way of doing this is of course the direct recording of the passage of vehicles. It might also be possible to accomplish something indirectly through cordon tolls at the edge of the congested area, in combination with specific parking controls. This, however, is likely to balkanize the area, providing an artificial separation of economic activity on the one side of the cordon from that on the other. Tolls at an adequate level limited to main thoroughfares might reduce this balkanization somewhat, but at the cost of producing severe "shun-pike" congestion on the parallel minor streets. Special license fees imposed on city street users, in the form of a flat monthly or annual charge, hardly meet the problem: it is difficult if not impossible, without rather elaborate checking mechanisms, to distinguish between occasional use by suburban and out-of-town vehicles, which could appropriately be exempted entirely, and regular commuting for which the maximum charge would be appropriate. The usual basis for special local license fees is the location where the vehicle is customarily garaged, which is almost entirely inappropriate. For example, a vehicle which is kept in an off-street parking facility privately paid for, and is used only for week-end excursions may be responsible for relatively little congestion, even making full allowance for the week-end parkway congestion.

Parking. Whether in connection with a cordon toll or in connection with a comprehensive electronic toll system, it will be desirable to have a closer check on the time at which cars enter and leave the traffic stream from parking space. Attended parking lots may presumably be required to provide this information in suitable form, possibly not without some resistance; on-street parking may at first appear to present a more formidable problem. The proposal formerly advanced by the writer for parking meters to be operated on a post-payment basis in which the rate of charge would be varied according to the number of parking spaces occupied and such that the deposit of coins would be required to recover the parker's key seems inadequate for this purpose.[7]

A proposal for curb parking control more in keeping with the above requirement is one in which a record is made of the time of parking and unparking as follows: for each local vehicle there is issued a parking card; on parking his car the operator inserts his card in the parking meter, presses a lever on the meter down against a spring, and withdraws a key belonging to the meter. This operation locks the lever down and the card inside the meter, where it can be inspected through a window; simultaneously a record is made of the time and the serial number of the card by imprinting an embossed serial number from the card and a time indication from a clock inside the meter through a carbon onto a paper strip. Subsequently, on the return of the operator, he reinserts the meter key, which releases the lever, causing a second imprinting of time to be made and the card returned to the operator. The resulting paper tapes are periodically collected, the imprinted data read photoelectrically and converted to magnetic tape, then processed on a computer, either separately or in conjunction with records made automatically as cars cross zone boundaries.

By feeding the parking data for an area into the computer first, the occupancy rates for each group of parking meters at various times could be ascertained, and on the basis of these occupancy rates appropriate charges could be assessed for each parking of a car; the data can then be sorted by car number and matched with data for the zone boundary crossings to produce appropriate charges for the trip from the zone boundary to the parking spot, or from one parking spot to another. All the charges for any given car can then be combined, modified where indicated by a factor relating to the nature of the car, combined with any outstanding charges and a bill sent to the owner, together with a card valid for the succeeding period for those not in arrears. Payment would be enforced in considerable measure by withholding of cards from delinquent owners: it could be made unlawful to operate a locally registered car in the area without a current parking card displayed appropriately in a holder on the windshield or elsewhere.

---

[7] William S. Vickrey, "The Economizing of Curb Parking Space," Traffic Engineering, November, 1954, pp. 62-65.

If corresponding records can be made for cars entering and leaving private parking lots and garages, and, perhaps more difficult, for cars parking in non-metered outlying areas, it might be possible to achieve a fairly efficient system of charges by combining these records with a record made of cars entering and leaving the metropolitan area at some cordon line, without having to equip cars with any sophisticated electronic or other apparatus. Cars coming off the street into a private facility would be able to register "off" by inserting the card in a receptacle similar to a parking meter, from which it could be retrieved only by again registering "on." This would enable all cars making trips fairly directly from one registration spot to another to be accounted for fairly accurately. There remains the problem of taxicabs, delivery trucks, and similar vehicles making devious and irregular trips. This could be taken care of, though somewhat awkwardly, by simply assuming that in the absence of evidence to the contrary, such vehicles registered as "on" the streets would be assumed to be circulating in the most congested areas subject to the highest rate of charge per minute, and providing opportunities in the outlying areas for such vehicles to register so as to qualify for an appropriately lower rate, e.g., by registering in an unused parking meter, or at extra registers provided for the purpose at convenient locations; it would only be necessary to register once for each more or less direct trip or trip segment beginning or ending in an outlying location.

There seems to be no easy way to deal with non-metered parking within the cordon, however, so that the necessity for having all parking within the cordon metered is likely to require that the cordon line be put rather closer in than is altogether desirable, both from the point of view of a reasonably complete coverage of significant congestion and from the point of view of locating the cordon itself at a point where the required operations would not themselves be a congestion factor. On the other hand with a little ingenuity it should be possible to avoid the necessity for "toll-plaza" operations if space is at all scarce. Outbound, it is the operator's advantage to register, so that a number of registration stations can be set up in tandem, with the driver being left free to register at any one of the stations. Inbound, while failure to register can be adequately penalized in those cases where the vehicle is going to "roost" somewhere within the cordon, there would be the possibility of evasion by those driving on through or making stops "on the run." But even here, since the operation is simply one of inserting the card in a slot, with no occasion for receipts or change requiring a human agent, it would be relatively easy to arrange for registers to be used in tandem, with signals to coordinate the sequence of use and indicate failure to register.

There remains the out-of-town vehicle. Where such vehicles are frequent visitors, it seems clear that they should be required to obtain cards on the same basis as local cars. For the occasional visitor or transient, this involves a relatively large administrative cost; moreover there is relatively little price elasticity to this particular category of use, and a considerable sentiment, arising both from commercial motives and traditions of "hospitality" in favor of generous treatment. Accordingly it seems entirely proper to arrange for such visitors to be given the "freedom of the city" insofar as street usage is concerned, to a suitably limited extent. This can be implemented by providing stations at a suitable distance outside the cordon where incoming visitors and occasional cars can obtain guest cards, valid for from two to ten days, showing conspicuously the license number of the car for which it was issued. A record is made of this license number when the card is issued, and this record is eventually checked to see that no more than the allowed number of guest cards are issued for any one car. It would probably be worth while to man the busier stations during hours of heavy traffic, particularly as they could be made to serve as general information booths as well; for light traffic routes and slack periods, the issuance of these cards could be automatic: for example, the applicant could be asked to write the license plate number through a plastic window and a single-use carbon on to the guest card, the carbon retaining the record; as the guest card is withdrawn, a transparent contact-adhesive film is pressed over the license number imprint to prevent further alteration. The guest card may also contain an embossed serial number, similar to those on regular cards, for further identification. Even though a guest card was issued as valid for, say, ten consecutive days, the subsequent processing of the records would show for how many days it was actually used, and it would be possible to specify that an occasional visitor was entitled, say, either to ten consecutive free days a year, or perhaps to six free days scattered through the year. It may be necessary to require the deposit of a dime for each card in order to prevent frivolous withdrawal of cards.

As compared with the cash post-payment type of parking meter the present charge account type has one major disadvantage in that the time lag in the feed-back of information to the vehicle operator as to how great the charge was for parking at a particular time and place is comparatively long, so that his response to a given level of charge may be considerably delayed. Major changes in patterns of vehicle use will probably take considerable time, however, in any case, so that this may not be as serious as it might seem at first glance. As compared with cash post-payment, the problem of the operator returning to the parked car with unsufficient change to pay the accumulated charge would be avoided. The meter mechanism should on the whole be less costly, particularly as there would be no direct interconnection among meters: assessment of the level of demand in the area would be by the computer program, not by interconnecting the meters; this would be not only a considerable saving in installation costs, but would make it possible to appraise the aggregate demand with much great flexibility and over a much wider area, so that the charges would not have to vary by such severe increments and a more stable equilibrium would be obtained. It would probably be necessary to make the release of the meter key conditional upon some pattern of holes or notches in the parking cards, otherwise there might be trouble from mischievous improper removal of meter keys; since there would be no gain to be obtained from such action, the keying need not be highly sophisticated and the meter mechanism can be kept reasonably simple.

Peak-off-peak transit pricing. The problem of suitable pricing of transit service to take account of the variations in cost between peak and off-peak service is relatively simple, at least for subway and commuter lines, and need not concern us here. Suitable automatic schemes for collection of sophisticated fare structures have been available, at least on paper, for some time. Recent developments in the direction of the use of subscriber cards and monthly billing (instead of cash payment, with or without refunds, as under former proposals) have been proposed for the San Francisco Bay Area Transit System, and seem well adapted to cases where the unit of sale is fairly large, as in systems that cater predominantly to medium-haul suburban traffic, as would be the case, for example, with railroad commuter service as distinct from local subway service. Also the clientele is of a character that would give rise to less difficulty in collecting on the basis of monthly bills. The principal unsolved problem is for local bus service, where lack of space and low levels of utilization militate against the use of complicated fare collection machinery, while any complication in the fare structure tends to overload the driver and slow service. The recent abolition of bus transfers in New York is a case in point, for while it does involve discriminations of a somewhat arbitrary sort between persons whose trips happen to be catered to by single long bus routes as compared to those having relatively shorter trips requiring the use of two different routes, it does significantly ease the burden on the operator.

An interesting suggestion has been put forth in connection with subscriber fare systems for suburban service: it would be at least conceivable that the agency furnishing the service should be set up as a membership organization along co-operative lines, and if so, one might argue that the subscribers should be entitled to deduct that part of their payments representing interest on capital and taxes, somewhat as the owners of co-operative apartments do. The deduction would have at least this much rationale: commuting to relatively low rent suburban housing is a more or less direct substitute for the payment of higher rentals in close-in housing. Indeed, one could present the further case that with the commuter service the efficiency of allocation of resources is being fostered rather than otherwise by the discrimination, in that the service is one offered under conditions of decreasing costs; this much cannot be said of the deductions for interest and taxes on dwellings. But this is admittedly stretching matters pretty far.

Financing the intra-marginal residue in urban transportation. While the most urgent need is to increase charges on rush-hour service of various kinds, especially motor vehicle usage, there will remain, ultimately, a fairly substantial portion of the cost of urban transportation that cannot be fully allocated on marginal cost principles. On equity grounds many would argue for allocating this residue as a charge added to marginal cost and paid by the users, and the second best principle would support this treatment at least to some extent. Thoroughness, however, requires that some attempt be made to trace out the way in which the assessment of transportation costs affects urban land use.

In this area more than most, extreme models produce paradoxical results. Land values are widely held to be created, or at least enhanced by transportation developments, particularly those that produce changes of mode or nodes in the transportation network. Indeed at times it almost seems as if it is imperfections in transportation that create land values. Where transportation has uniform costs, models seem to indicate that the better transportation is, the lower land values will be, as for example with the assertion often made that if transportation could be made instantaneous and costless, site value would disappear. Even with less extreme models, if demand for urban land as a whole is considered to be completely inelastic, and if rents are made to vary so that for all developed urban land the sum of rents and the transportation costs of the activities carried out on the land to and from the center of the model are constant, then land rents are proportional to total transportation costs, and cutting transportation costs in half will preserve the same geographical pattern of activity with rents likewise cut in half.

A considerably different model emerges if we admit some elasticity into the demand function for access to the urban center, the amount of access being measured by the area of land occupied and the price of access being the sum of land rent plus the cost of a uniform amount of transportation per unit of land from the site to the center. If for simplicity we assume that all land uses have the same transport-to-the-center requirement, and that the demand function is unity, then the total price paid in terms of rent plus transportation cost is constant, and of this total price one-third is rent and two-thirds is transportation cost. The total cost can be thought of as a cylinder of height $h = tr$, where r is the radius of the developed area, and t the transportation cost per unit distance; at the edge of the cylinder transport cost is tr and rent is zero; at the center transport cost is zero and rent is tr; total rent can be represented by the cone fitting into the cylinder, which will have one-third of its volume. As the transport cost t increases, r shrinks (as the cube root of the transport cost rate), rents at the center rise, but the developed area shrinks, so that outside the developed area rents fall to zero. With a more elastic demand curve for access to the center, total rents would increase as transportation costs fall, but rents at the center would still fall, the increase being accounted for by enlargement of the developed area.

In order to produce a model in which improvement in transportation raises rents at the center, it would be necessary to produce a demand curve for access to the center which is in a sense perverse, i.e., in which the larger the aggregate, the more a given buyer is willing to pay for access to the center, and in which this effect is so strong as to outweigh the diminishing marginal utility of access to a center of a given size to successively less eager buyers. In other words if the Nth renter is willing to pay R for access to a center of size N, then even though the N + 1st renter would be willing to pay only $R - \delta$ for access to a center of size N, the Nth renter is willing to pay $R + \delta + \epsilon$ for access to a center of size N + 1, and the N + 1st renter is thus willing to pay $R + \epsilon$ for access to a center of size N + 1. Such an upward sloping demand curve would indeed imply that a decrease in transportation costs would increase the marginal value of access to the center, and hence increase rents throughout the developed area. Thus it is possible for an improvement in uniform transportation to increase property values at the center; but whether the economies of scale in urban aggregations of the present size of most of our large cities are still significant enough to bring about this result would seem to be unlikely on an impressionistic basis.

On balance it seems likely that improvements in uniform transportation, as exemplified by travel in private automobiles over a standardized network of city streets, would benefit owners of outlying property more than owners of centrally located property, and that if intra-marginal highway costs are to be charged against property values at all, they should be charged primarily against peripheral property values rather than against central property values. Indeed some of the deterioration observed in downtown areas over a period coinciding with the growth of the automobile relative to transit's decline is rather suggestive of the agreement of reality with some of the above models; this result may not be any adventitious one resulting from strangulation by inadequate development of facilities, but may be an inherent property of a non-focusing transportation system.

Models reflecting node effects. Models which will reflect the focusing effect of a transportation system with pronounced nodes are rather hard to come by, and it seems reasonable to suspect that a number of other facts, such as the variation in the relative demand of activities for space and for transportation, the possibility of linkages that do not pass through the center, the possibility of creating space through building upward, and the effect of the aggregate magnitude of activity on the demand for space would all interact with the existence of transportation nodes in affecting the pattern. To illustrate the lines on which analysis might proceed, I venture to suggest the following outline for a model, though without necessarily implying that the model can be worked through satisfactorily short of using successive approximations on a computer.

1. All transportation takes place along the lines of a rectangular grid, which however is dense enough to allow one to neglect the spacing between transportation lines. Initially the cost of transportation is uniform per unit of distance, with the result that the developed area takes the form of a tilted square, the diagonals of which are two of the transportation routes at right angles to each other. In subsequent stages improvements are made in transportation facilities along these two main axes, so that the marginal cost of travel along them is reduced as compared to travel off these axes. The result is to spread the developed area out in a four-pointed star.
2. A unit of activity is defined as an activity originating and terminating one unit of traffic. For simplicity the amounts of traffic originated and terminated are assumed to be equal. Activities vary according to some simple distribution as to the amount of space they require per unit of activity.
3. The frequency with which a unit of transportation is carried on between any two units of activity is inversely proportional to some power of the cost of the transportation (the so-called "gravity" model).
4. Space can be provided at a given location by constructing buildings of varying height and lot coverage, at (linearly) increasing marginal cost of additional space as the ratio of space to ground area increases.

Specifying the conditions of equilibrium is a fairly straightforward matter, buildings being constructed to a height at which space rent equals marginal cost, the difference between marginal and average cost appearing as ground rent, which, on this basis, is proportional to the square of the building height. Activities presumably locate in such a way as to minimize the sum of rent and transportation cost; there is, to be sure, some question at this point whether the transportation cost considered to be paid by an activity should be the one-way transportation cost or the two-way transportation cost: in principle it would be the two-way cost, but one might want to consider contexts in which it is more likely that an activity bears only the costs of outgoing transportation and incoming transportation is fully prepaid by the shipper, or vice versa. Presumably, also, each activity would want to minimize not the cost of reaching a given set of correspondents by transportation, but the cost of transportation as adjusted for such shifts of correspondents as would be induced by the application of the gravity rule.

I have not had time to develop this program fully so I merely offer it as a suggestion to anyone who may be possessed of the computer programming facilities and mathematical skills required. Without going into the matter further a guess at the results is risky, yet one might hazard the guess that significant differences would emerge between the patterns that result when transportation costs are uniform as compared to what they are when transportation along certain channels is specially favored. If this expectation is substantiated this would be a further reason for using property tax revenues to subsidize nodecreating transit rather than unfocused motor vehicle transportation.

Rationales for subsidy of one mode by another. If for one reason or another outside financing cannot be obtained to meet the intra-marginal residues of transportation as a whole, there still may be sound justification for financing the intra-marginal residues of mass transit by the levying of charges on urban motor traffic, even if these charges cannot be made to vary closely with the diurnal and other variations in congestion. In one sense, such a use of motor vehicle user revenues for the finance of transit would be merely the equalization of the subsidy that now is given to the urban motorist, partly in the financing of actual outlays on streets out of property taxes, and partly in the failure to account at all for the rental value of the space the motorist occupies.

In some cases a politically more acceptable rationale can be derived from adventitious historical circumstances, as when it is proposed to turn some of the San Francisco Bay Bridge toll revenues over to the Bay Area Transit Authority, on the ground that originally the bridge did carry rail transit cars of the Key System, and the tracks were later removed to make way for additional roadways. The plea is made that this diversion of tolls is in lieu of the recapture of the bridge space for rail transit use. Actually, while use of the bridge would be considerably cheaper in capital cost than the present plan to construct a new tunnel for the transit system under the bay, use of the bridge would be rather less satisfactory because of the awkward approaches that would be required and the severe speed restrictions that seem inevitably to apply to rail equipment operating over suspension bridges.

A more basic argument for subsidy of this sort can be derived from the following example, however, which illustrates how far off the "every tub on its own bottom" philosophy can get when misapplied to what seem superficially to be aggregates of similar tubs (even without an introduction of decreasing costs!). Suppose a facility of type M attracts rush-hour and nonrush-hour passengers in the ratio of 1 to 4, and costs $1.00 for each rush-hour passenger provided for and $.20 for each nonrush-hour passenger. On the other hand a facility of type T costs only $.75 for each rush-hour passenger and $.15 for each nonrush-hour passenger, or uniformly 25 per cent less; however, it attracts only one nonrush-hour passenger for each rush-hour passenger. If no differentiation between rush-hour rates and nonrush-hour rates is possible, facility M can break even with a charge of [$1.00 + 4($.20)]/5 = $1.80/5 = $.36; for facility T the break-even fare is [75¢ + 15¢]/2 = 90¢/2 = 45¢. Thus with each facility required to be self-liquidating, the passenger is offered a 9-cent fare differential in favor of facility M, alike in the rush hour and in the nonrush hour, whereas the consequence of his choosing facility M rather than T is to increase the costs, by 25 cents in the rush hour and 5 cents in the nonrush hour.

Of course, if it were true that if five rush-hour and five nonrush-hour passengers were to shift from T to M, they would automatically convert themselves in some way into two rush-hour and eight nonrush-hour travellers, then the 9-cent reduction in fare that they would obtain would be justified. But there is no reason to expect any effect in this direction, and an effect of even a fraction of this magnitude would be a highly unlikely occurrence. Actually, if peak-off-peak differentiation in charges is impossible for one reason or another, the next best thing would be to reverse the relationship between the charges, and raise the charges on M to 45 cents and lower the charges on T to 35 cents, resulting in a subsidy of about 25 per cent to T from excess revenues of M. If for the M facility we read streets and highways and for T we read suburban rail service, the correspondence with the typical facts is reasonably close. Thus if charges for peak use are ruled out, there is an even stronger case than would exist otherwise for subsidizing mass transportation at the expense of vehicular traffic.

Police and Custodial Service

Police expenditures amount to slightly over 10 per cent of city expenditures, ranking below education and sanitation, but are perhaps less amenable than most other services to financing by means of specific charges. There is nevertheless something to be gained by an examination of the ways in which the revenue structure can be brought more closely in line with the costs of performing this service.

Unfortunately, not very much is as yet known concerning the specific factors that influence or should influence the level of police service provided. In New York City, under a "Post Hazard Plan" promulgated in 1955 and subsequently revised, an index of the relative volume of police problems in the various areas is used as an aid in allocating personnel. The index uses the following items with corresponding weights: Crimes of personal violence, .25; other crimes and offences, .20; juvenile delinquency, .15; accidents and aid cases, .10; population, .10; area, .05; business establishments, .05; school and recreation areas and crossings; .05; and radio alarms transmitted, .05. Obviously, at most .20 of the total weight is assigned to factors that could be made the basis for some form of tax, and even here the basis is impressionistic rather than based on any rigorous study.

There would seem to be a relatively close relationship between the characteristics of buildings in an area and the magnitude of the policing problem. But even if this can be substantiated, it is not clear how this relationship could be converted into a tax base and whether it would be desirable to do so if it could be done. Buildings converted to single room occupancy may appear to increase the policing problem, but it is at least possible that a tax on such occupancy would only result in the tax being passed on very largely to tenants. Indeed, to the extent that police problems are most intense in low income neighborhoods, it may be almost inevitable that any attempt to levy a tax in proportion to some factor that seems to be causally related to policing costs will result in a severely regressive levy. This can be illustrated, for example, if it is proposed to allow some sort of tax credit for the provision of doormen, full-time janitors, and the like, whose presence would tend to reduce the policing problem.

There is, indeed, a whole array of situations where the line between private and public policing is unclear. At one time it was the practice in New York for city police to be detailed as guards to accompany payrolls and other similar transfers. This had serious ill-effects: gratuities were often offered in connection with this service, to the detriment of morale; the gratuities failed in most cases to cover the full cost, so that in effect an incentive for converting to a payment-by-check basis that should have been brought to bear was held off. A further effect was the diversion of unduly large numbers of police to this duty at peak periods near week-ends.

In some instances police are hired to perform such duties during their off-work hours; this also may lead to abuses. Similar difficulties occur in conjunction with sporting events and other occasions where large numbers of persons are assembled. The range over which this problem of drawing the line between that for which the public police department will be responsible and that which is a private responsibility is a wide one. At one extreme, the Morningside Heights Association has recently arranged for a corps of private police to patrol the area as a supplement to the city police. At the other there is said to be at least one case where the municipal police deliver the morning newspapers as a part of their early morning patrol. While this could obviously cause difficulty if carried too far, there are obvious complementaries between patrolling an area and performing other functions that may be worth taking advantage of, and of course if police are involved in such activities, the possibilities for obtaining revenues by appropriate charges should not be overlooked. But on the whole there is at this stage little that can be said definitely about appropriate modifications of revenue structures on the basis of their impact on police expenditures.

Recreational Facilities

Expenditures on recreational facilities account for about 5 per cent of general expenditures, or about one-third as much as highways (in terms of cash outlay). In terms of the nature of the benefit, one could argue that here is an even stronger case for the levying of specific charges. However, it is rather more difficult to isolate a marginal cost that makes very much sense because in many cases the anemity provided by a park is enjoyed by occupants of abutting property, even if they impinge in no way upon the enjoyment of others. Some uses are joint rather than exclusive, in that one goes to an event in part in order to be a part of the multitude. Even where the use is definitely exclusive, as in playing golf or tennis in periods of heavy demand, and accordingly efficient allocation would definitely call for the levying of a specific charge sufficient to equate demand and supply, there is a case to be made on distributional grounds for rationing by queue rather than by price. Many groups have more time than money, and it can be argued that it is desirable to preserve a reasonably wide variety of areas in which those who possess little coin of the realm can cash in that coin of which they have relative abundance.

Thus while a professor of economics, accustomed to value time highly and to think of queuing as an essentially wasteful process may at first take a dim view of a system where people line up for hours to get on a public golf course, this may not accurately reflect the feelings of those who do the lining up. Existence of a queue, to be sure, is *prima facie* evidence of inefficiency, in that if somehow reservations could be handed out to approximately the same group of persons who eventually play, the waiting, at least, could be eliminated. The cost involved in handling the reservations, however, may be greater than that of the waiting (this is essentially the analog of the former situation with respect to vehicular traffic, where it was maintained that the cost of toll collection would be greater than the cost of the congestion it was to eliminate). Permitting non-transferable reservations to be made on a first-come, first-served basis involves an alternate waste of induced excessive pre-commitment to a specific plan of behavior some time in advance, on pain of forfeiting the privilege represented by the reservation. Again at first glance one might suppose that this waste could be eliminated by making the reservations transferable, to the mutual benefits of transferor and transferee, but this would only make the reservations either an attractive area for speculators to operate in, or require the application of some prior criteria for the issuing of the reservations.

Where the capacity of the facility is fairly rigid, as with tennis courts, and the demand somewhat unpredictable, because of the influence of the weather if for no other reason, it will in any case be almost impossible to clear the market with any degree of precision through pricing alone, and in such cases a combination of pricing and queuing is likely to occur. In the absence of distributional considerations, a proper balance is to be sought between the wastes of queuing and the wastes of underutilization of the facilities: the higher the price the less the queuing but the greater the underutilization in periods of unusually low demand. On the other hand the lower the price, the lower the revenues over the range of prices that is relevant here, while the greater, presumably, is the favorable effect on the distribution of income.

The outcome of the balance of these considerations is not a cut-and-dried matter, in any particular case. In considering the weight to be given to the losses of queuing, however, it is appropriate to consider the degree of compulsion involved in the queuing: if the associated service is one easily dispensed with or for which there are reasonably close substitutes available, as for example where private facilities are available at moderate fees, then the queuing losses can be assigned a relatively low value in the appraisal of alternatives, whereas if the queuing is more nearly associated with a necessity, with only relatively remote substitutes available, the costs of the queuing are likely to weight more heavily in the decision.

The same principles apply to the case where there is a temptation to continue the same fee that is inadequate to eliminate queues in peak periods over into periods of slack demand where the facilities are lightly used. In these cases the argument seems to be that the demand is highly inelastic over the range of prices in question anyhow, and that the redistributive effect of reducing the fees is less likely to be favorable, partly because of the absence of the selective effect of the queue, and partly on the ground that possession of leisure during the off-peak period is evidence of economic prosperity. There is also the problem of whether in view of the relatively low level of demand the facility should be closed down entirely during off peak periods, or whether, contrariwise, only the fee collecting element should be closed down. This is not an important problem, but it is one that often seems to have been resolved more in terms of administrative convenience than in terms of any serious economic appraisal of the situation.

Education

Education is by far the largest single item in local government budgets, and there is no dearth of material on its financing. This is not the place to attempt a review of the volumes that have been written, but rather to sketch out new avenues of approach that might follow from an examination of the economics of the problem.

If provision of a uniform minimum standard of educational opportunity for all were more nearly a fact, rather than the rather remote aspiration that it actually is, the problem might be considered relatively simple. In fact, communities differ widely in their ability as well as in their willingness to provide a high grade of education. While intrastate differences in ability are met to a moderate degree by state aid formulae, the degree to which interstate differences in ability are met by federal aid is as yet minimal. The desire to provide adequate or even superior educational facilities thus often is severely constrained by the lack of adequate public revenue resources.

The problem is aggravated to a certain degree by the imbalance of internal migration. Regions that are net exporters of educated personnel fail to realize the tax base that would ordinarily result from the activities of the persons they educate, a process that has certain vicious circle elements in it as areas with high educational aspirations relative to their tax base fail thereby to attract the base necessary to support the aspirations without recourse to burdensome rates. It is easy to exaggerate the importance of this, but it seems clear that some means of breaking through this situation is needed. The essence of the problem is that the opportunity for a profitable investment in education is being missed because the parents of the children involved may be unable to finance the education, either individually or collectively, while the community finds it very difficult to finance an investment from which the returns will accrue elsewhere. The problem is how to arrange for the repayment to the investing community of some of the returns generated by that investment in education.

The somewhat exotic proposal that these considerations seem to point to is as follows. Each federal income taxpayer should be required to report on his income tax form the state (or school district) in which he received his public education, if any. A portion of his tax would then be turned over to the state (or school district) so indicated. The intended effect would be primarily one of making "educational export" regions more willing to upgrade their educational standards in view of an expectation that even though the region might not benefit directly from the better education of those who leave, the region would nevertheless get a return in this form.

Such a principle is of course capable of great modification in detail. It could, for example, be applied only prospectively, to taxpayers in the future that are now getting educated; presumably the incentive to educational export regions to upgrade their education would remain as strong, even though they might have to borrow to do it on the strength of the expectation of these revenues. They merely would not obtain an unexpected windfall from the incomes of their previous students. However, this would be subject to the perennial problem of the inability of any government to commit itself effectively to a program that would have to extend for so many years into the future if it were to be effective at all, and in the absence of some mechanism of effective commitment, the incentive effect on current expenditures on schools might not be realized. Also, it would be possible to limit the distributions in some way to those educational systems that represent a high degree of effort in relation to the revenue resources available. Another possibility would be to vary the amount of the income tax payable by the individual according to the degree to which his income might be considered to be the result of superior or inferior educational facilities, the variation in tax above the minimum level being the amount available for redistribution back to the educational agencies.

The difficulties of putting any such plan into practice are obvious and probably insurmountable, and accordingly it must be considered as being presented as an aid and stimulus to discussion rather than as a serious proposal. Possibly a more likely, but less effective, variant would be to propose that some of the federal grants in aid to the states might be calculated on the basis of their being, in effect, ex post compensation to the exporting states for the loss of human capital they experienced as a result of net emigration in the past, insofar as this capital export could be considered to reflect the value of the education given. Another way of doing it would be to make current and future migration patterns the basis for grants, without attempting to relate the grants to the income actually yielded by the investment in education.

Health and Hospital Services

The main element that is peculiar to health and hospital services is the close interrelationship between municipally financed services and services covered by insurance, and the presence of a "moral hazard" element, by which is meant the tendency to make unjustified use of the facilities because they are either free or covered by insurance. In a sense the "moral hazard" here is only an acute form of the general tendency to overuse a facility that is underpriced; if a difference exists it is mainly that here the justification for use is supposed to reside more in objective medical facts and less in individual preferences, and the moral hazard arises because the objective medical facts are never quite as objective as the concept of a welfare standard or an insurance indemnity would like them to be. The moral hazard arises from many motives, from the doctor's desire to have the patient where it is easy to visit him, or the use of the hospital as a refuge, or to the desire to take advantage of section 105(d) of the federal income tax, which allows a deduction for sick pay for the first seven days of illness only if the employee is hospitalized during that time. The problems are so diverse that about all that could be done here is to simply list the area as one in which there is the possibility of some financing by fees.

Public Utility Services

Although such public utility services as electricity supply, telephone service, gas distribution, mail service, parcel delivery, and even newspaper delivery are not ordinarily included among the services considered as part of the standard pattern of governmental activity at the local level, the lines between these services and those such as water supply, garbage collection, and sewage disposal, that more typically are so considered, are to a degree arbitrary. Of these, the one that is most frequently added to the list of municipal activities, electricity supply, is perhaps the one for which the special powers of the municipality are least needed.

Dissatisfaction with the results of public utility regulation of private utilities, plus the attractions of the availability of relatively low cost capital and certain other tax advantages relative to federal and state taxes are among the major factors which have led to the entry of municipalities into the business of supplying electricity, but these factors are common to the other utilities as well. In some cases the establishment of a municipal electricity service is the result of such special stimuli as the TVA and other public power agencies. But to a large extent the fact that these forces were effective with respect to electricity but not to other utility services can be laid to the fact that of all of the predominantly privately supplied ones, electricity is the one that is most nearly an absolute necessity for the typical urban resident. Actually, it is precisely this characteristic that would make it possible for a privately operated utility to come reasonably close to an optimum allocation of resources through the adoption of a schedule of charges that will closely reflect marginal cost and yet at the same time extract a sufficient additional revenue from the intra-marginal consumption to earn a normal return on its investment. There are several ways of doing this, but the most appropriate would be to charge rates for kilowatt-hours at the appropriate marginal cost, and assess in addition a front-foot charge to cover the basic cost of the distribution system. The front-foot charge could be expressed, if need be, in the form of a higher rate per kilowatt-hour for the first x kilowatt-hours per month per front foot; since practically all consumers would be using more than this initial block, the result would be essentially marginal cost pricing. This device is available to the electric power company primarily because the use of electricity is so nearly universal that no customers to speak of would find it advantageous to refuse service because of this initial rate.

With nearly all of the other services, however, an attempt to charge for the service according to front feet, to cover the basic costs of traversing the streets so as to cover the area effectively, would result in many potential customers refusing the service in order to escape the cost, and it does not seem likely that a privately operated service would be empowered to assess charges on property owners who do not take the service. It is here that there is a definite opportunity for improvement in the efficiency of allocation of resources through the collection by the municipality of a special frontage tax to defray the basic traversal costs of these various services, and then make available to the residents these various services at rates that would be fairly close to the incremental cost of rendering the service. It would not be absolutely necessary for the municipality to render the service on its own account: this could be done by contract, though of course in this case care would have to be taken lest abuses occur, and if anything of this sort were to be done at all, many would be inclined to favor some method that would not result in the subdivision of responsibility and the introduction of opposing monopolistic interests. Including mail delivery in this list is perhaps a bit quixotic in view of the firm preemption of this field by the federal government, yet it is technologically no more unreasonable to contemplate the local handling of mail that is transmitted nationwide by a national service than it is to contemplate local distribution of electric power generated and transmitted over a wide area by a federal agency.

SPECIFIC FINANCING AND THE OVER-ALL FISCAL PICTURE

In considering the extent to which specific charges should be used in each instance, some attention must of course be given to the relation of the costs and revenues in that area to the over-all fiscal picture of the governmental unit, both as to the aggregate amount and as to the distributional impact of the costs and revenues. The first of these elements can be adequately represented for most purposes by introducing as a parameter the "marginal cost of public funds."

The Marginal Cost of Public Funds

The marginal cost of public funds is defined as the net reduction in the over-all allocational efficiency of the economy resulting from the raising of an increment of revenue from a given source, expressed as a ratio or percentage of that increment in revenue. It includes, over all, the loss in consumers' surplus, the loss in producers' surplus or economic rent, the marginal costs of administration, and the marginal costs of compliance. If we are considering an established set of taxes and charges and only the rates are under consideration, the compliance and administrative costs may be approximately constant, so that they can be neglected, and we can focus our attention on the effects of the rate changes on surplus.

The simplest case to analyze is that of a simple excise tax on a commodity produced under conditions of constant or increasing costs. In Figure 1, AC is the supply curve, BC is the demand curve, C is the price and quantity resulting under competition with no tax, and ACB is the net social gain generated as a result of the production of the commodity in question, divided into consumers' surplus CNB and producers' surplus or rent, ANC. If a tax equal to DE is imposed, per unit of output, the supply curve inclusive of tax moves from ACS to A'ES' there is a tax revenue of EDGF, the consumers' surplus is reduced to EFB, producers' surplus is reduced to AGD, and there is a net loss of potential surplus to the community as a whole of DEC, representing the excess of the value VECW of the output that is no longer being produced VW, over what it would cost to produce it, VDCW.

Suppose that now the tax is increased to HI. Consumers' surplus is now IJB, revenue is now HIJK, and producers' surplus is now AKH, the total loss of potential surplus now having increased to CHI. The net increase in this loss, EDHI, is what is to be compared to the increase in revenue, which is HIJK - DEFG. This comparison can be made in terms of the somewhat simplified Figure 2, in which we in effect regard the tax as the margin which the government acting as a monopolist adds to what it pays for the total supply to determine the price at which the total amount will be sold. BC is the demand curve as before, and BQ is the corresponding marginal revenue curve (in the case of straight line demand curves it is twice as steep as BE), representing the net increase in gross revenues from the sale of one more unit at the correspondingly lower price. Similarly, AR is the marginal cost of purchase curve, representing the cost of buying one more unit, allowing for the fact that buying one more unit will raise the price which must be paid for all units in the open market. The difference RQ is then the loss in net revenue resulting from the purchase of one more unit at an over-all additional cost of VR and its sale at an over-all additional revenue of VQ; this is what is to be compared to the loss of net social surplus represented by ED, the difference between the value VE the buyer would place on the additional unit and the cost of producing it, VD.

A third way of presenting the situation is to subtract the supply curve AC vertically from the demand curve BC to get the curve BC in Figure 3 which shows the relation between the tax rate and the amount sold. This can be regarded as the net demand for the government's (costless) services in transferring the product from sellers to buyers. The total surplus generated if there is no tax is ABC, as before, and the amount that is sacrificed if a tax of DE is imposed is DEC, the total tax revenue being ADEF. The marginal government revenue curve BUT being the marginal curve

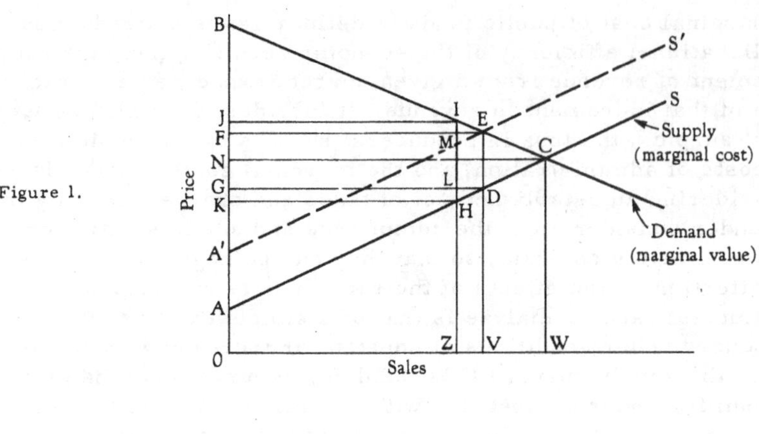

Figure 1.

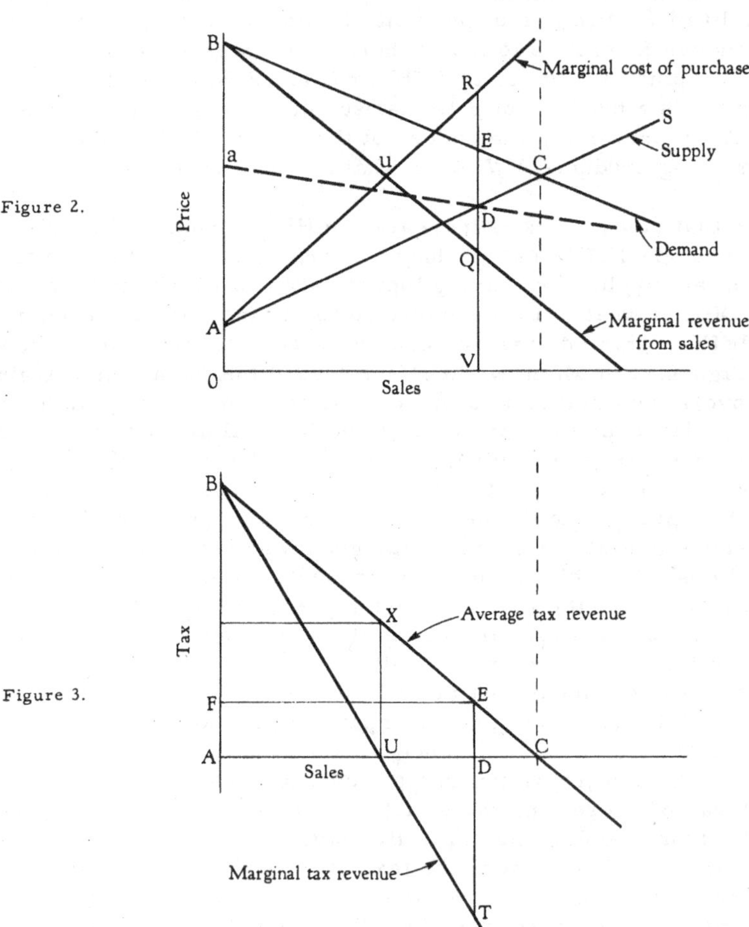

Figure 2.

Figure 3.

corresponding to the average revenue or tax rate curve BEC, represents the net revenue from arranging for the sale of an additional unit. Increasing the tax to reduce the sales by one unit thus increases revenues by DT (the negative marginal revenue), and increases the net social loss by DE. It is the ratio ED to DT that is the marginal cost of public funds; this obviously begins at zero for no tax at C and increases steadily as the tax rate is increased, becoming infinite as the tax is raised to the point UX, where the revenue is a maximum.

If taxes are being raised from several such sources at once, the optimum arrangement is one in which the ratio ED/DT is the same in all cases. It can be shown that for excise taxes this relation obtains if the ratio of tax to price for each commodity taxed is proportional to the sum of the (absolute) inelasticities of demand and of supply, provided that we measure the inelasticity of demand net of the tax (i.e., as though the demand curve had been shifted downward by the amount of the tax; inelasticity is the reciprocal of the elasticity).

It is tempting to apply this analysis directly to the case where the government is itself the supplier. This is correct if the supply is obtainable at constant costs, but in the case where the service is subject to increasing or decreasing costs it is necessary to allow for the fact that the government is in effect the recipient of the producers' surplus that in the excise tax case accrued to private individuals, at least in the more usual case where the increasing costs of the government service arise within the service itself and is not a result of pushing the prices of factors of production on the market up. Thus in Figure 2, the loss from reducing the consumption by one unit as a result of raising the price is still ED, but the accompanying change in revenue is the saving in cost, indicated by the marginal cost curve which would be the supply curve in the former case, or VD, less the reduction in revenue, indicated by the marginal revenue curve VQ. Thus in this case the marginal cost of obtaining public funds by raising the price is ED/QD. The "tax" DE, in this case representing the excess of the price over marginal cost, is, as a percentage of the price VD, to be made proportional only to the inelasticity of demand, the inelasticity of supply for this purpose being taken as zero. The same analysis applies if the service is supplied under conditions of decreasing cost, as would be indicated by the dotted curve aD.

Marginal Redistributional Coefficient

Similarly it would be nice to have a coefficient of redistribution that could be applied, to indicate more or less how much misallocation the community was willing to tolerate for the sake of how much redistribution. In principle, it would be just as inconsistent to have two alternative systems which show an equal marginal cost of public funds but a widely varying marginal impact on distribution. Distribution, however, is not something that can easily be reduced to a single parameter, and in addition opinions vary so widely on the desirability of income redistribution that anything that could be done along this line would be highly subjective. In view of these difficulties, the matter is left with this brief mention.

SUMMARY

Enough has been said to indicate that even though the decision to provide a given facility or service may well require a weighing of costs and benefits, benefit may not provide a suitable basis for the assessment of charges to defray the costs. The decisions made by governments in deciding whether to offer a service and of what extent and character, and those made by individuals in availing themselves of whatever service is provided are of a different scope and often of a different dimension, so that the comparisons that are pertinent for the one decision are not always relevant for the other. Generally speaking, the considerations adduced in considering appropriate charges lean rather more towards the taxation of land rather than improvements, not necessarily in terms of its market value, but rather in terms of such parameters as area and frontage. Even where benefits seem to be most closely measurable in terms of improvements, the causal nexus is found to be more nearly independent of actual improvements and to relate more to the fundamental site-determined characteristics. Finally, a comprehensive approach to the pricing of municipal services may lay the groundwork for a new and more fruitful approach to over-all urban planning in terms of economic costs as well as of architectural design.

5

Toward Quantitative Evaluation of Urban Services

by Russell L. Ackoff[1]

Many of those who have worked on planning problems--whether they have dealt with a city or a nation, whether they have been qualitatively or quantitatively oriented--have been aware of the difficulty of obtaining a satisfactory evaluative measure of the services provided by a community. Without such a measure it is extremely difficult, if not impossible, to evaluate alternative plans in an objective way.

Current and future work in the development of mathematical models of urban phenomena will make it possible to predict more accurately the consequences of any plan. This is a large step forward but without a rational criterion for evaluating these consequences, such models could lead to the degradation rather than the development of our communities.

Those planners who have no way of determining what are the values of a community are likely to impose their own values on the community. As Britton Harris has observed:

> Most unfortunately, we find a certain number of planners, whose income is adequate and whose physical needs are well satisfied, who are completely incapable of projecting themselves into the situation and value structures of those less fortunate or simply different from themselves; they place at the forefront of their preoccupation those values which they are now striving to satisfy and neglect those which they have already met. More honorable but equally unrealistic are those who have made a too-educated choice and who try to impose their pursuit of the good life on others.[2]

---

[1]I am greatly indebted to the following for their critical evaluation of earlier versions of this paper: Stafford Beer, C. West Churchman, John W. Dyckman, Britton Harris, Erik Johnsen, Harvey S. Perloff, William J. Platt, Gordon D. Shellard, and Lowdon Wingo, Jr.

[2]Britton Harris, "Plan or Projection," The Journal of the American Institute of Planners, No. 26 (1960), p. 267.

In the last two decades we have learned a good deal about the construction of planning-decision models in military and industrial contexts. As yet, relatively little such modeling has been done on governmental and community planning decisions. Much of what we have learned of decision-modeling in the armed forces and industry, however, is applicable to planning of the public sector of society.

Ideally, models of planning decisions should have the following form:

$$V = f(C_i, U_j) ,$$

where

V = a measure of the utility (value) of an urban service: alternatively called the "outcome" or "performance" variable or "objective function."

C_i = the controlled (planning) variables: those which are subject to control by the decision makers.

U_j = the uncontrolled variables which affect the utility of the service rendered.

f = the mathematical function which expresses the way in which the controlled and uncontrolled variables affect the utility of the service.

Normally, such a representation of a decision situation must be supplemented by a series of statements which express the limitations or restrictions on the values of the controlled variables. For example, a controlled variable such as "annual construction expenditures in a community" may have a limit imposed on it by the community's budget. Such constraints can normally be expressed by a supplementary set of equations and inequations (i.e., statements involving the relations "is greater than" or "is less than").

The regional growth and transportation models which have been, or are currently being constructed are not in the form I have just described. Their outcome variable is very seldom a measure of the utility of the services rendered; rather they use such descriptive (as contrasted with evaluative) variables as "number (or density) of households at a specified time." Hence, even after such models have been used to predict the consequences of alternative plans, the evaluative problem remains.

The task which I have set for myself in this paper is to determine the feasibility of constructing satisfactory measures of over-all performance of an urban service. Many economists have worked on such measures and methods for obtaining them, but--except for recent efforts at cost-benefit analysis--most of their efforts have been differently oriented than is this one. Most of them have attempted to obtain measures without reference to a <u>decision model</u> in the sense in which I use this term. Consequently, they tend to look for gross characterizations of urban services.

I am concerned here with revealing the components that go into a gross (macro) measure so that decision models can be constructed which relate controlled and uncontrolled variables to the over-all measure through component (micro) measures. Revelation of these micro measures is necessary for determining what variables ought to be included in a decision model. For example, we cannot construct a production-control model for minimizing total production cost in industry, even where we can measure total production cost, unless we can identify and measure such components as the cost of carrying inventory, production setup costs, costs of labor and raw materials, and so on.

In addition, the approach I shall take here allows and encourages us to take noneconomic aspects of the service into account.

If we have a suitable measure and model, a major worry of many economists is removed, a worry that derives from the difference in being able to measure the total value of a service and the marginal change in this value produced by changes in the service. A decision model yields a measure of value for any state of the system and hence provides a basis for calculating not only total value but the marginal values of changes from any one state to another. What is more important, it makes optimization possible; that is, it permits us to determine how to set the controlled variables so as to obtain the best over-all performance of the system. Once a decision model has been constructed, either by mathematical analysis or simulation, we can obtain a set of values of the controlled variables in the form

$$C_1 = g_1(U_j)$$

$$C_2 = g_2(U_j)$$

$$\cdot$$

$$\cdot$$

$$\cdot$$

such that the measure of performance, V, can be maximized (or minimized, if appropriate) for any values of the uncontrolled variables.

I should like to emphasize the fact that no matter how important evaluative measures of performance are, they are by no means all that is required for objective rational planning procedures. They are of only limited value without decision models. Unfortunately, I cannot consider here how decision models should be constructed so as to yield either optimal allocation of resources to services, or optimal utilization of resources which have been allocated to a service.

Those planners who are discouraged by difficulty in research and those who feel that their expertise lies in their ability to make good qualitative evaluations and judgments, may be inclined to assert that quantitative evaluation of urban services is impossible. They might argue that too many "intangibles" are involved. But to argue in this way is to fly in the face of history which has shown that in each age the intangibles of preceding ages are made tangible. It is virtually impossible to prove that a concept cannot be measured in any way. Consequently, it is much more constructive to leave the question of measurability open until measurement has actually been accomplished.

Another argument which is sometimes given against measurement of the utility of urban services is that even if it were accomplished, it would not be used because political decision makers base their decisions on other than community interests. Assuming politicians do act in this way, it would certainly do no harm and it is likely to do some good to demonstrate objectively to the public that the politician's choices do not best serve its interests.

Measurement of a previously unmeasured concept is seldom accomplished in one fell swoop. Measures of concepts, like their meanings, evolve slowly; but since successful attempts are based on unsuccessful ones, I can draw some comfort from the fact that my efforts may yield a foundation on which future successes may be built. If planning recommendations can be reached in such a way as to carry with them the prestige of science, they are much more likely to be considered seriously than they are at present. There is good evidence of this in the impact on underdeveloped countries of scientifically-based national planning. Of course there is much more to "selling" planning recommendations than this, but the value of a label, "made by scientific labor," should not be ignored.

In the section that follows I shall consider in general terms some of the methodological requirements for constructing a measure of the utility of an urban service. Then I shall try to construct measures of the utility of urban transportation and educational services. I have selected these services not because I have any special knowledge of them--it will be apparent that I do not--but because my friends in planning feel they represent the range of difficulties that might be encountered in developing measures of services and because they involve larger expenditures than any other services. Finally, I shall try to draw from these two efforts at formulating evaluative measures some methodological conclusions that apply to other services.

METHODOLOGY OF QUANTITATIVE EVALUATION

This discussion of methodology first develops the concept of a "best decision" in very general decision-theoretic terms and then considers problems that are specific to the context of social planning.

Planning problems always involve decisions concerning the "inputs" and "outputs" of a system. By an "input" I refer to any valued resource (e.g., money and time) or state (e.g., comfort) which is consumed or dissipated in connection with obtaining goods or services. The goods, services, or state produced or obtained are the "outputs"; for example, automobiles, transportation, and good health. In other words, inputs are what must be sacrificed in order to obtain the outputs.

The terms "cost-benefit" may be used in place of "input-output" but in so doing it must be remembered that their connotation is not constrained to economic considerations.

"Input" and "output" are relative concepts in the sense that the output of one activity may be the input of another. For example, money is spent (input) to obtain an automobile (output) which is then used (input) to get from one place to another (output), and so on. Furthermore, inputs and outputs in a particular process may be the same kind of thing. For example, money may be invested in order to make money. In short, what we take as inputs and outputs depends on how we define the system in relation to its environment. Inputs and outputs are not natural phenomena; they are conceptual constructs.

The relativity of the concepts "input-output" does not destroy or limit their usefulness. In fact, this relativity is a great advantage because it allows us to apply the same concepts to sub-systems and to the systems that contain them, to long- and short-range problems, and thus to relate their analyses.

The evaluation of a process always involves a comparison of its inputs and outputs. There are essentially three different ways in which this can be done and, hence, there are three different kinds of measure of the effectiveness of a service.

1. The amount of input required to obtain a specified output
   For example, the "cost per unit" may be used to evaluate a production process, "cost per ton mile" to evaluate a freight system, and "time to run a mile" to evaluate track men.

2. The amount of output yielded by a specified amount of input
   Examples of this kind of measure are "number of units produced per day" or "per dollar," "number of miles travelled per hour," and "number of problems correctly solved in a specified time."

3. The difference between (or ratio of) the amounts of output and input
   Output minus input is "net return" or "profit." I use these terms in a very general way: they are not restricted to monetary inputs and outputs. Similarly, the ratio of output to input is "return on investment."

I shall refer to these measures as (1) the input measure, (2) the output measure, and (3) the net return or profit measure.

Those measures that have been applied to urban services have generally been of the first or second type, not the third. The reason is that the profit measure requires that inputs and outputs be measured on a common scale, and this is frequently difficult to do. A profit measure, however, has an important advantage over input and output measures which can only be used to compare alternatives, but not to determine whether any one is good enough. The profit measure can do both tasks. For example, one manufacturing process may be less costly than another, but both may be too costly to yield a profit at the going price that can be obtained for the product.

Ideally, then, we would like to construct a measure of profit of a service, but in practice it may be necessary to settle for an input or an output measure. If this is the case, it is important to remember that the question, "Is the service good enough?", must be answered on qualitative grounds.

Inputs and outputs are related to individual and social objectives. Inputs involve those objectives which are preservative in nature; that is, objectives concerned with the maintenance or retention of something of value: keeping what one has and wants to keep. Outputs, on the other hand, are related to objectives which are acquisitive in nature, that is, which involve obtaining what one does not have and wants. Consequently, in any type of problem situation it is necessary to take three steps if one is to "set up" for developing a measure of the utility of the solution.

1. Identify the individuals and collectives (groups, organizations, corporate bodies, etc.) which are involved in the activity under study.

2. Identify those preservative and acquisitive objectives of the participants which are affected by the relevant activity.

3. Identify the inputs and outputs associated with the relevant objectives.

Identification of the participants can usually be accomplished by an examination and analysis of the activity involved. Identification of relevant objectives and associated inputs and outputs is likely to be more difficult, but it raises no problems which are unique to urban services and which have not been studied extensively in other contexts. These problems are discussed in the paper by Lichfield and Margolis which appears in this volume (pp. 118-146).

Once the first three steps have been taken, construction of the measure can begin. It involves three additional steps, of which the first two are:

4. Provide an operational definition of each relevant input and output, and specify how these are to be measured and what scales are to be employed; and

5. If more than one scale is specified in step 4, select one of these into which the others are to be transformed. We will call this the "effectiveness" scale.

For example, one input variable may be <u>cost</u> expressed on a monetary scale and the other may be <u>time</u> measured in minutes. If measures of these two are to be added, they must be made commensurate, either minutes must be transformed into dollars, dollars into minutes, or both into a third scale. In most cases it is easiest to use a monetary scale of effectiveness because money is the most common and universal standard which is used in evaluating goods and services.

Transformations may be based on either objective or subjective data. Objective transformations of inputs or outputs which are measured on one scale are obtainable when these inputs or outputs are converted behavioristically into ones which can be measured along another scale. For example, delays in delivering packages (measured in time) to customers of a department store (an output of a delivery service) in some cases can be shown to decrease the future purchases (measurable in dollars) of these customers. Hence the relationship between lengths of delay and profit can be determined objectively. Objective transformations then are based on study of the overt behavior of the relevant entity.

Subjective transformations are obtained when the decision maker is asked to select equally-valued points on two scales (e.g., to plot the transformation function). Here opinion is used.[3]

Once all the necessary transformations have been obtained, a single measure for the relationship between input and output can also be obtained as a measure of the system's effectiveness. But this may not be enough. Assume that effectiveness is measured on a dollar scale. Assume further that effectiveness of a particular service can be described by a single

---

[3]For a detailed discussion of these techniques, see R. L. Ackoff, <u>Scientific Method: Optimizing Applied Research Decisions</u> (New York: John Wiley & Sons, 1962), Chapter 3.

point on this scale. Then it might seem that the best service plan is one that yields the greatest effectiveness. But this is true only if the utility of units on the effectiveness scale does not decrease with increases in the number of the unit. Put more precisely, let n represent a certain number of units on the effectiveness scale, and V(n) represent the utility of this number of units. Then in order to use a point on an effectiveness scale as the ultimate measure of value of a service, it is necessary that

$$V(n) < V(n + 1)$$

for every value of n that might be obtained by a service.

This is clearly not always the case. For example, the value of increasing the number of street lights or traffic signal lights per mile is not necessarily positive. As a minimum, then, it is necessary to establish the validity of the inequation given above.

Furthermore, the effectiveness of an urban service system does not remain constant but varies as a consequence of the instability of some of the relevant uncontrolled variables. Therefore, performance of the system should be characterized by a distribution of measures of effectiveness, the mean of which is its "expected effectiveness." Again, however, a system with maximum expected effectiveness is best only if the inequality, $V(n) < V(n + 1)$, holds.

Now we can formulate the last step in the process of constructing a measure of the utility of a service.

6. Find the utility function of the effectiveness scale.

A number of difficult methodological problems are associated with deriving such a function. Several of the more important of these are discussed in the next section.

Since the purpose of measuring the performance of services is to improve the allocation of resources to them, and the utilization of these resources, the performance of different services (e.g., transportation and education) must ultimately be expressed on the same scale of effectiveness. Only if this is done can the relative values of investments in different services be determined.

To sum up, in order to provide a quantitative evaluation of a service we must identify the participants, their relevant objectives and the associated inputs and outputs, for each of which a suitable scale should be specified. If more than one scale is used, these should be transformed into some one effectiveness scale (usually monetary) and, finally, the utility function of the effectiveness scale should be constructed.

Obtaining a Social Utility Function for an Urban Service

The basic questions to be answered in developing a measure of the social utility of a public service seems to me to be the following:

1. How should a benefit which is derived from a combination of services be allocated to those services?

2. Which of a chain of benefits that is initiated by rendering a particular service should be attributed to that service?

3. How should a measure of the social utility of a service be aggregated out of measures of the utility of that service to individuals?

Any one of these questions requires more space than I can give it here, but I will try to suggest the form of *an* (not *the*) answer to each.

Allocation of a benefit among interacting services. Consider two services, S_1 and S_2, which may be used separately or together (in sequence or simultaneously). Suppose each one alone produces an outcome, O_1 and O_2, respectively; and that in combination they yield an additional outcome, O_3. Let C_1 and C_2 represent the costs associated with each, and B_1, B_2, and B_3 represent the benefits associated with the outcomes. We can represent the *independent* value of S_1 (i.e., S_1 without S_2) as

$$V(S_1 \text{ without } S_2) = V(B_1 - C_1)$$

and the independent value of S_2 as

$$V(S_2 \text{ without } S_1) = V(B_2 - C_2).$$

If both services are used together we get the *interdependent* value

$$V(S_1 \text{ with } S_2) = V(B_1 + B_2 + B_3 - C_1 - C_2).$$

There may also be a joint cost, C_3, involved. The important point is that

$$V(S_1 \text{ with } S_2) = V(S_1) + V(S_2)$$

because S_1 and S_2 are not completely independent. We cannot attribute the value of B_3 to either service separately nor divide it among them because both are *necessary* and neither is *sufficient* for B_3. The general principle involved here is that a benefit should be attributed only to that service or set of services which is sufficient for producing that benefit in the environment of interest.[4]

When we are concerned with possible modifications of several services (say, S_1,S_2 and S_3), multiple *levels* of value are involved:

1. The independent values created by each service $V(S_1)$, $V(S_2)$, and $V(S_3)$,

2. the interdependent values created by first-order interactions: $V(S_1S_2)$, $V(S_1S_3)$, and $V(S_2S_3)$,

3. the interdependent values created by the second-order interaction: $V(S_1S_2S_3)$.

---

[4] For a detailed discussion of ways of handling nonindependent utilities, see P. C. Fishburn, *A Normative Theory of Decision Under Risk*, Ph.D. thesis, Case Institute of Technology, Cleveland, Ohio, 1961.

The total value yielded by the three services is

$$V(S_1 \text{ with } S_2 \text{ with } S_3) = V(S_1) + V(S_2) + V(S_3) + V(S_1S_2) + V(S_1S_3) + V(S_2S_3) + V(S_1S_2S_3).$$

If n services are involved, there are n^{-1} levels of interaction.

It should be observed that most of the "externalities" of interest to urban economists are due to interactions of services.

If a new recreational facility (S_1) is built by a city, certain benefits (B_1) will normally follow. If, in addition, transportation services (S_2) to the facility are improved, there will be additional recreational benefits (B_3) which are the joint product of the improved recreational and transportation services. If the transportation services alone were provided it might produce its own independent benefits (B_2). The value of the transportation system is enhanced by the recreational facilities and the value of the recreational facilities is enhanced by the transportation services.

The method of evaluating transportation services which I will develop later measures $V(S_2) + V(S_1S_2)$,[5] the independent and the dependent value of the service. But if the same method is applied to the recreational services it will measure $V(S_1) + V(S_1S_2)$. We should not attribute $V(S_1S_2)$ to two different services and hence do "double accounting." Such a procedure does give us the value of the transportation system, given the recreational services, and the value of the recreational services, given the transportation services; but part of each of these values is the value that is jointly produced. This does not present a difficulty, however, because the purpose of evaluating an urban service is to determine how to change it so as to maximize increases of value to the community without regard to how the service and its environment interact to produce the increase in value. For example, if the new recreational facility (S_1) is already available, and we consider improving transportation services to it (S_2), then the net gain of so doing is $V(S_2) + V(S_1S_2)$. Although $V(S_1S_2)$ should not be attributed to S_2 alone, it can be attributed to the decision to provide S_2. This distinction is an important one because planners are more concerned with evaluating decisions to modify services than with measuring the value of services already in existence.

When concentrating on one service, we must remember that if the costs or benefits associated with any interacting services are changed, the dependent value of the service of interest will most likely change. In principle, if we can measure $V(S_1) + V(S_1S_2)$, then if we change S_2 and measure this sum again, the change in its value must occur in the interaction term $V(S_1S_2)$ and is due to the change in S_2.

Allocating from chains of services and benefits. Starting with any event we can always trace back to obtain as long a tree of prerequisite events as we have the patience to construct. For example, suppose A's life is saved by a doctor in the public health service. The doctor could not have rendered the service unless he had received a medical education. He

---

[5]It is possible to modify the procedure to estimate $V(S_1S_2)$ alone and, hence, to also obtain a separate estimate of $V(S_1)$.

could not have received the relevant medical knowledge in school had it not been discovered by some earlier researcher, and so on back to Adam and Eve. Clearly, we cannot attribute the values derived from current services to an original "first cause." We need some way of truncating the tree. A reasonable way of doing so is to allocate a value received now in a particular environment to those services which are sufficient in that environment for producing the value. This means that we attribute the value of saving A's life to the relevant services in his current environment. If we want to go further and attribute the value of the public health service to something we will have to go back to the most recent set of conditions which were sufficient for making it possible.

The same principle applies when examining the progression of benefits that emanate from a current service. Suppose A receives funds from a public welfare agency. He gives some of it to B who buys something that benefits C, and so on. The benefits derived by B and C depend on A's action and not on that of the public welfare service alone. Therefore, we attribute to the current service only those values for which it is sufficient, not those for which it is only necessary.

If the service consists of transporting A to a place of work where he is employed by B, the value to B of A's being at work cannot be attributed to transport, because it depends on what A does at work. If, on the other hand, A is travelling at B's expense and on B's time, B will have values which are relevant to transport.

Aggregating measures of individual values. Suppose we can determine the maximum amount of money that each member of a community is willing to pay for a public service. For each individual this would be the monetary equivalent of the value he places on the service. However, we cannot add these monetary values to obtain an aggregated measure of value to the community because the same amount of money may represent different amounts of value to different people. For example, if both a pauper and a millionaire are willing to pay the same maximum amount of money for a service, we would conclude that the pauper values it more highly.

This difficulty would disappear if we could measure the value of a service on an "interpersonal" or "absolute" scale of value, but no such scale is available yet.[6] We can measure the relative values of alternative services or even different amounts of money to an individual.[7] Can measures of relative value of a service to different individuals be made commensurable?

Some economists have argued that interpersonal comparisons of value are impossible and, hence, rational social decisions (i.e., ones that maximize expected social utility) are also impossible because social values cannot be aggregated from individual values on any rational basis. The principal reason given is that any aggregation of values of individuals requires assumptions of a political or ethical character; and that the validity of such assumptions cannot be established by science, and hence must be

---

[6] I believe such a scale is possible in principle and have suggested what it might look like. See R. L. Ackoff, "On a Science of Ethics," Philosophical and Phenomenological Research, No. 9 (1949), pp. 663-72.

[7] For techniques of so doing, see J. von Neumann and O. Morgenstern, Theory of Games and Economic Behaviour, 3rd edition (Princeton: Princeton University Press, 1953); Ackoff, op. cit. (footnote 3); and Fishburn, op. cit. (footnote 4).

"irrational." I do not accept this position but obviously it is too complex an issue to argue here.[8]

The method of aggregating individual values which I suggest is based on the observation that the value which an individual places on an amount of money depends on how much money he has available for spending. This in turn suggests that the value of an amount of money is some function of the ratio of that amount to the total amount of money that the individual has at his disposal (i.e., disposable personal income or some other related quantity). The difficulty with using this ratio as a measure of the relative value of an amount of money is that it assumes a linear relationship between amount of money and its value. There is an abundance of experimental data to show that this is not the case; the value of money tends to increase approximately as the square root of the amount. Therefore, we can modify use of the ratio as follows: Assign the value "1" to the individual's disposable personal income and using this as a base, determine the relative value of the maximum amount of money that an individual is willing to pay for the service under study. Individual measures of this type can be aggregated by simple addition to obtain a measure of social value. It is apparent that the sum of these relative values of the things purchased with an income will usually exceed 1.0. This is to be expected: the value of the goods and services purchased with an income should exceed the value of the income. Otherwise, why earn and spend it? This allows for the reasonable expectation that a specified income can yield different amounts of value, depending on how it is spent.

If two or more persons have one disposable income between them and if we treat the income as a unit then we must multiply the resulting value of the service by the number of people dependent on that income. If an income is divided among two or more people for independent disposition, each part and individual should be treated separately.

If the egalitarian assumption that underlies the assignment of a value of "1" to each person's disposable income is not satisfactory to the researcher, he may try several alternatives and compare the outcome (i.e., the impacts on planning). Since we do not currently have an objective ethical basis for aggregation, the planner could do worse than "go to the public" with the issue.

It is important to realize that the necessity of some ethical assumption is not avoided by using macroscopic economic measures as indices of social value. Such measures (e.g., gross national product or standard-of-living indices) always involve an aggregation of individual measures, usually simple addition of amounts of money. The assumptions which are required to justify such an aggregation are much more difficult to justify, I believe, than are the assumptions which I have made here. The availability of macro-economic measures does not itself justify their use as measures of social value.

In the discussion of educational services which appears below I will examine some of the assumptions which are implicitly made in using macroscopic economic measures.

---

[8] I have argued it in another place. See Ackoff, op. cit. (footnote 3). Churchman has done so also in what I think is the most convincing way yet published. (See C. W. Churchman, Prediction and Optimal Decision, Englewood Cliffs, N.J.: Prentice-Hall, 1961.) Our argument asserts that although ethical assumptions are necessary, they are subject to scientific investigation.

Henceforth, when I refer to the value of any input or output, I refer to its relative value to an individual which is determined by using his disposable income as a base with a value of "1" assigned to it.

EVALUATION OF TRANSPORTATION SERVICES

The term "transportation" generally means the use of any public or private conveyance to move persons or goods from one place to another. Such conveyances may include bicycles, horses with or without wagons, automobiles, street cars, helicopters, pipe lines (for water), and so on.

I should not like to restrict the term to uses of conveyances so that walking can be included, since sidewalks and pedestrians must be taken into account in the design of any transportation system. Furthermore, the inclusion of walking avoids having to distinguish between users and non-users of the transportation system since all but a few completely immobilized people in a community will make some direct use of the system so conceived.

By "transportation services" I refer to (a) provision by the community of facilities, equipment, and personnel used to make transportation possible, and (b) legal or administrative action affecting the provision of such facilities, equipment, and personnel by private individuals or groups (e.g., the effect of public policy on the number, size, and location of privately operated parking lots). The services include construction, maintenance, and policing of sidewalks and roads; operation of public conveyances; and any related activities of public agencies.

Participants

The participants in the system fall into two main groups: operators and users. I will not consider the operators here because their interests in the system do not involve transportation as such, but rather the characteristics of their jobs. These interests, of course, would have to be taken into account by administrators of these agencies, but they have to be considered by planners only to the extent that they affect the cost, quality, and kind of services rendered.

Relevant Objectives

What input objectives are relevant to users of a public service? Put another way, what resources, valued opportunities, capabilities, or desirable states may have to be sacrificed in order to obtain the service? It seems to me that all possible sacrifices can be put into one or more of the following classes:

> The valued goods or money that the user must barter or pay for the service: <u>the cost</u>.
>
> The opportunity to do other things he would rather do with the time occupied by the service: <u>lost time</u>
>
> The ability to do and enjoy things in the future; that is, he may suffer <u>harm</u>.

The effort that is expended in receiving the service, effort that he would rather expend for other purposes: lost energy.

The pleasure deriving from his physical and mental state while receiving the service: loss of enjoyment.

These types of sacrifice may be incorporated into the following input objectives:

To minimize costs. Relevant costs appear to be of two types: those paid indirectly by the individual through the community as a whole, and those paid directly by the individuals using the service. The costs to the community include construction, purchasing, installation, maintenance, and operating costs such as those for policing the roads and operating public conveyances. Costs to the users involve either direct outlay of fares or the cost of using private vehicles, including parking and tolls. "Cost" is used here in the restricted monetary sense.

To minimize lost time. Most people want to get to where they are going as quickly as possible because travel prevents them from doing something they would prefer to be doing. But where this is not the case (e.g., when working or reading on a plane or train, or conversing or sight-seeing in a car), total time is not critical. Hence it is lost time, not total time that is critical.

To minimize harm (i.e., to maximize safety). People do not want to expose themselves or their property to damage or destruction.

To minimize lost energy. People do not want to arrive at their destinations too de-energized to engage in the activity for which they went there. They may be willing to expend some but not most of their energy.

To maximize enjoyment. Most people want to avoid exposure to unpleasant sensations (discomfort) while in transit: tactile (e.g., bumpy or hot rides), visual (e.g., dirty or ugly views), auditory (e.g., noise), and olefactory (e.g., offensive odors). Increase of such displeasure as a result of transport is an input. On the other hand, pleasure derived from the trip can be considered as negative inputs, or outputs.

It is one thing to identify these input objectives but another to define them and measure the degree to which they are attained. I will reconsider each one in turn, try to construct a relevant scale for each and convert it into a monetary scale which, in turn, can be used to obtain a quantitative evaluation of transportation services to the community.

Scaling Attainment of Input Objectives

Cost. Defining and measuring the annual monetary cost to the community (C_1) of providing transportation services may involve considerable complexity. Even though we may be able to identify each of the component costs, there are usually difficulties in accounting for overhead and depreciation of equipment and facilities, and discounting current expenditures that affect future services. Such problems have received attention from experts in public fiscal policy, but they nevertheless persist. Planners can point to situations in which the choice between alternative decisions depends on the discount rate used. Clearly, if the choice between alternatives does not depend on this rate, it has no significance. However, if this choice does make a difference and there is no apparent rational basis for selecting an appropriate discount rate, the principles developed in decision theory for "uncertainty" decisions are applicable. A sample of "reasonable" discount rates can be selected and use can be made of that one which yields a minimax, Hurwicz, or minimax regret solution to the problem.[9] The same approach can be taken to other financial questions, separately or in combination.

The direct cost of trips to the users (C_2) does not seem to present any great conceptual problems. There may, however, be some difficulty in collecting the information which is necessary for estimating these costs, but a great deal of experience has been gained in recent years in collecting good trip data and relevant economic information.

Lost Time. The time used in transporting people and in transporting goods must be treated separately. Since goods in transit cannot normally be used or worked on in any way, all of its transit time is usually considered to be lost. The cost of this time is commonly calculated by many companies and presents no major problem; it is a function of the time in transit, the cost of the goods being moved, and the cost of an idle investment per unit time. This in-transit inventory cost is generally relatively small.

The time people spend in transit is another matter. Recall that it is only the lost time whose monetary value must be determined. Practical considerations seem to restrict us to two approaches to estimating the monetary value of lost time per trip: (1) we can estimate the monetary value to an individual of his lost time to those whose time it is, or (2) we can estimate the potential economic value of the total travel time to the community as a whole and divide this by the number of trips taken. Both approaches are incomplete: the first omits values of lost time to "others," and the second considers only economic value of total (not lost) time. Where possible, I think we should use both approaches and use the larger estimate obtained, since both methods yield underestimates. My personal preference is for the first approach because I think it is likely to yield less error and because it could be extended to include the value to "others" if it were practical to do the necessary amount of work. These difficulties seem to me to be of a lower order than those involved in extending communally oriented economic analyses to cover noneconomic values.

Returning to the monetary value of time, then, the question is: What monetary value do people place on the time they lose in travel? For the person who is travelling on "his own time," I suggest using his own value of that time. For the person who is travelling on "someone else's time," I suggest also using the value of the person whose time it is (i.e., who pays for or controls it).

---

[9]For details, see R. D. Luce and H. Raiffa, Games and Decisions (New York: John Wiley & Sons, 1957).

It is possible to avoid the complexity of finding out how people would spend the time freed from travel without ignoring its significance. Suppose we could present an individual with an alternative A to a means of transport B which we want to study so that A and B are alike in all respects except time and cost of travel. Suppose further we could manipulate the price of the faster of the two so that we could determine the largest amount of money that the individual is willing to pay for it. If we could repeat this for a number of different travel times we would get a curve something like the one shown in Figure 1. If such a curve could be obtained for each user of a mode of transport for each trip on that mode, we could estimate by extrapolation the weighted monetary value of the time lost in a trip (C_3).

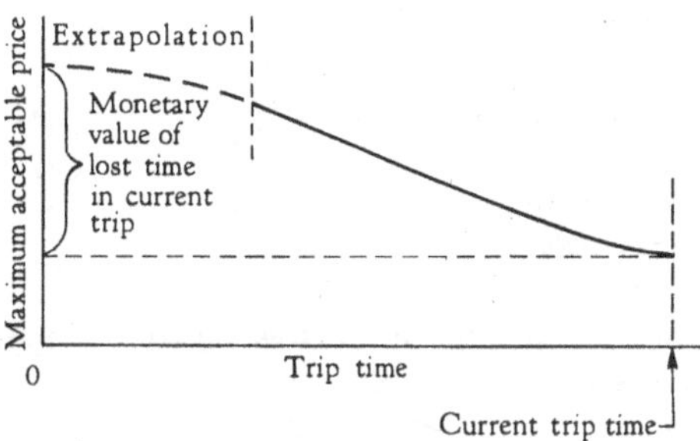

Figure 1. Illustrative response to decreased time and increased charges.

Now, clearly, from a practical point of view we cannot obtain such a curve for each trip taken, but we could do so for a sample of trips. We might proceed as follows:

1. For two means of transport alike in all respects except travel time, increase the charge for the faster of the two by one increment at a time. Note the percentage of original users selecting each alternative at each charge.

2. Repeat step (1) for a sample of reduced transit times.

3. From the data obtained from steps (1) and (2) and a knowledge of the number of total trips taken, an estimate of average monetary value placed on time lost per trip on the normal mode of travel could be obtained.

An experiment of this type could most easily be conducted where express and local services are provided on public transport, or where parallel bridges, tunnels, or toll roads operate for private transport.

The appropriate design of such an experiment and the development of the appropriate estimation procedures are not as simple as the brevity of my discussion might suggest, but I see no reason why such an experiment is not feasible. We are already performing similar manipulations of travellers in some modes of transport, for example, in the application of additional charges for jet flights.

An economic approach to the value of time spent in travel would involve determining some such quantity as the earnings that could be obtained with this time or the gross or net economic value of the product that it could yield. In such an approach we would have a great deal of difficulty in determining how much time is lost, and we would ignore noneconomic values. The noneconomic values may very well be the most important ones involved, particularly in a society in which there is no shortage of labor and in which reduction of work time is an important objective.

Safety. The calculation of the monetary value of damage to, or destruction of property has occupied a good deal of attention of actuaries and economists. The difficulties it presents are of a lower order than those involving damage to or destruction of life.

Since World War II a number of studies have been directed toward an economic evaluation of human life. According to Rosch[10] these have dealt with the following concepts:

1. Cost of a life: the cost technically necessary to save a life.

2. Price of a life: the expenditure that the community is willing to make in practice to save a human life.

3. Compensation for death: cash award or compensatory pension to near relatives.

4. Cost of a man: aggregate expenditure on consumption, investment, and public services which are devoted to him.

5. Product of a man: the value of his crude production (contribution to Gross National Product).

6. Loss of a man: the loss that a death imposes on the community.

Discussion of these measures will be found also in Thédié[11] and Dessus[12]. Injuries may be evaluated in the same way.

It should be noted that the price of a life which is what we are interested in here must on the average be less than the difference between its product (output) and its cost (input).

Unfortunately, there has been little study of how much monetary value an individual places on his own life. One is likely to assume that he places an infinite value on it, but this is clearly not the case. We all expose ourselves to danger in order to obtain something else that we want. Such risking of life is evidence that we place a finite value on it. The amount of risk which an individual is willing to incur provides a basis for obtaining his

---

[10]Dr. Rosch, "De Quelques Prolongements et Ramifications en Divers Domaines," Revue Française de Recherche Opérationelle, No. 5, 1961, pp. 125-38.

[11]G. Thédié, "Report of Discussion in Session on Economic Decisions that could Involve Loss of Life," in The Proceedings of the Second International Conference on Operational Research, J. Banbury and J. Maitland, eds. (London: English Universities Press, 1961), pp. 784-86; and C. Abraham and J. Thédié, "Le Prix d'une Vie Humaine dans les Décisions Économiques," Revue Française de Recherche Opérationelle, No. 4, 1960, pp. 157-68.

[12]G. Dessus, "De l'ineluctable Measure des Incommensurables et de ce qui Peut s'ensuivre," Revue Française de Recherche Opérationelle, No. 5, 1961, pp. 139-56.

monetary evaluation of his life. Assume two services alike in all respects except the probability of surviving them. Then, the difference between the maximum amounts that the individual is willing to pay for these two services divided by the increased probability of survival is his monetary evaluation of his life, assuming he knows the relevant probabilities. For example, if the increase in probability of survival is .001 and he is willing to pay $100 for this, he places a value of 100/.001, or $100,000 on his life. Once we know the dollar-value that a person places on his life we can determine the monetary value of the risk to life by multiplying the dollar-value by the probability of survival. Similar computations can be made for other types of harm. The total cost of risk to safety per trip (C_4) would be the sum of these values.

Energy. Techniques exist for measuring the amount of energy expended in human activity. We can determine, for example, how much more energy is required to walk up a set of stairs than to use a parallel escalator. If we place these side by side and adjust the escalator's speed to match that of walking, then the largest amount an individual is willing to pay for use of the escalator is his monetary value of the energy saved. By extension of the idea behind this simple test, experiments could be designed to obtain individuals' monetary evaluations of their energy which is expended in an activity (C_5).

Enjoyment. First let us deal with the question of physical comfort. Comfort is a state which is influenced by a large number of physical characteristics of the environment. Although some of the characteristics (e.g., temperature, humidity, illumination, and smoothness of ride) can be measured, measurement of comfort itself is much more difficult. I think this will eventually be done, but we need an alternative approach until such measurements are developed.

Suppose we construct a classification of qualitatively defined states of comfort by various combinations of the physical factors which affect it. The monetary value of such states could be determined experimentally in much the same way as I have suggested evaluating other inputs. The maximum amount a traveller is willing to pay for additional comfort is the monetary value he places on it. The cost of discomfort for a particular trip (C_6) is the difference between the maximum price that the trip-taker is willing to pay for the most comfortable trip possible and the maximum price that he is willing to pay for this trip.

Experimentation. In order to obtain the kinds of estimates of the monetary value placed on time, safety, energy, and enjoyment which I have suggested, experimentation with actual transportation facilities is required. Such experimentation is long overdue. There has been a hesitancy on the part of operators of transportation facilities and planners to experiment on their users, but the fact is that the short-run inconveniences that such experimentation might cause are less serious than the kinds of inconvenience to which users are frequently subjected by "normal" breakdowns and delays caused by weather, strikes, and so on. The additional short-run costs to the users, if any, would be more than compensated for by the long-run gains that would be made possible by use of the information that such experimentation would yield.

The various classes of travel provided in rail and air travel show the possibility of varying different input requirements of travel. The kinds of experiments called for here would require little or no more manipulation of users than is required by current systems.

I believe these considerations show that planners should develop a considerable experimental capability. Drawing boards and social surveys are not enough. If planning is to become scientific it has no alternative but to become experimental.

Although I have discussed experimentation with respect to each input variable, a separate experiment is not required for each one. Modern techniques of experimental design make it possible to obtain the necessary measurements from a single set of experiments.[13]

What I have tried to show up to this point is that input objectives can be formulated in such a way that estimates of the degree to which each one is obtained can be formulated on a common scale, a monetary one. There is nothing sacred or necessary about this scale; it is simply convenient because our economic system already requires of each consumer that he evaluate alternative kinds of services and determine how much he is willing to spend for them. I am suggesting, however, that the planner make these evaluations in a more controlled and comprehensive way than does the average consumer.

Outputs

Ideally, we would like to measure the value of the output of a transportation system on the same scale as is used to measure input; in this case, on the dollar scale. The value of the output of such a system, however, depends on the nature of many other services and facilities provided in the community. For example; the value of transportation depends on the geographic distribution of stores, schools, places of work, recreational facilities, and so on. The value of each of these, in turn, depends on their accessibility. It is clear, therefore, that the value of transportation cannot be determined in the abstract; it must be determined in a specific environment. Comparative evaluation of transportation systems in different communities can be accomplished only if the effect of all other relevant differences can be cancelled out.

If we could measure the output of a transportation system along any well-defined scale other than dollars, it would be possible to use an input or output measure of value. For example, suppose we measure the output of a transportation system as "man-miles per year." (For the moment I am overlooking transport of goods which could be characterized by "ton-miles per year.") With such a measure we appear to be able to evaluate different systems by determining which one yields the least total (generalized) cost per man-mile, or one which maximizes the man-miles per unit cost. But this is not the case for systems in different communities. As I have indicated, differences in geography and facilities impose different requirements on a transportation system and, hence, unless the effects of these differences can be cancelled out, the systems cannot be compared. Perhaps this is not too serious because the concern of the planner is with alternative systems for the same urban area. But here we run into another kind of difficulty.

Let us compare two transportation systems, A and B. If the costs of A and B are the same, but B yields more man-miles and this has value to

---

[13]For example, see W. G. Cochran and G. M. Cox, Experimental Designs, 2nd edition (New York: John Wiley & Sons, 1957).

the community, then system B is the better. Certainly if B costs less, in addition, it is the better system. But suppose B costs more and yields more man-miles, then we cannot determine which system is the better unless we can express in dollars the value of the difference in man-miles.

More important, unless we can obtain a profit measure by subtracting input from output, we cannot express the value of transportation services on the same scale as the value of other services and, hence, cannot use these measures as a basis for the rational allocation of resources to a variety of urban services.

This is not to say that if we cannot obtain a profit measure we ought not to bother with an input or output measure. As in all scientific effort, we should try to come as close to the ideal as possible because with each successive step the domain in which qualitative judgments are required is diminished, and hence the magnitude of error is likely to be reduced. By proceeding in this way we can also make explicit what questions are being answered subjectively or qualitatively, and by such exposure we tend to accelerate their subsequent investigation. Keeping this in mind, let us turn to the possibility of transforming the output measure (man-miles and ton-miles per year) into one that is monetary.

One might reasonably argue (as the economists have for many years and as I have above) that the monetary value of a service to a person is the maximum amount of money (C_{max}) that person will pay for it. We expect that for some increase in the cost of transportation each person would begin to eliminate less important trips. As the cost increased further, the least important of the remaining trips would be eliminated. As the cost continued to increase the person would either (a) move to a dwelling which required less or no use of transportation services, (b) change his place of work to one that is accessible without transportation, or (c) move to another community. In any of these cases his mileage would tend to zero. Corresponding reaction could be expected from those who transport goods within the urban area.

Now let us combine the various measures we have discussed into a measure of over-all performance of a transportation system. Recall the following symbols:

C_1 = annual cost of transport system to the community (indirect costs)
C_2 = direct cost per trip to trip-taker
C_3 = monetary equivalent of lost time per trip
C_4 = monetary equivalent of risk to safety per trip
C_5 = monetary equivalent of expended energy per trip
C_6 = monetary equivalent of loss of comfort per trip
C_{max} = maximum price at which a trip would be taken

Then the value of the trip-taker's inputs is given by

$$V(\text{Inputs}) = V(C_2 + C_3 + C_4 + C_5 + C_6) = V(\sum_{2}^{6} C_j)$$

The value of the output to the trip-taker is

$$V(\text{Output}) = V(C_{max} + C_3 + C_4 + C_5 + C_6) = V(C_{max} + \sum_{3}^{6} C_j)$$

Consequently, the net-value received for a trip is

$$NV = V(\text{Output}) - V(\text{Input}) = V(C_{max} + \sum_{3}^{6} C_j) - V(\sum_{2}^{6} C_j)$$

The social value received over a year from transport would be the sum of the NV's over all trips taken that year. This sum divided by the annual cost to the community (C_1) gives the social value received per year per dollar spent by the community on transportation services. The same kind of measure of performance can be obtained for each mode of transport or for geographically defined subsystems. Such measures can be used to allocate community resources to modes and subsystems in an optimal way; that is, in a way which maximizes the increase in the social value per year per dollar spent. Similarly, if such measures were obtained for each type of public service it would be possible to obtain an optimal allocation of community resources among and within these services.

Now I turn to another type of community service, education.

EVALUATION OF EDUCATIONAL SERVICES

In my efforts to construct a measure of the value of the output of educational services I ran into a great deal of trouble. The output seems more complex and intangible than any of the other services I considered, including a number that are not discussed here. The difficulty persisted and did not seem to dissipate with reflection. An examination of the relevant literature showed that others had found this same difficulty. Most of those who have given the problem some attention, like Groves, attribute the difficulty to the fact that

> Education is a consumption good, as well as a productive factor; it is an end in itself as well as a means to other ends. In this role it aims to provide the facility for enhancing the enjoyment of life and wisdom to choose among competing values.[14]

The impression one gets from these discussions is that the *extrinsic* value of education might conceivably be measured, but its *intrinsic* value probably cannot be, at least at the present time. Although it seems clear to me that education has a value which falls outside the realm of economics, it is not clear what these other values are and how they can be measured.

Education provides the information and skills necessary for the continuation and expansion of scientific effort, for economic development, and for health. It attempts to produce well-adjusted, ethical, and moral individuals. It creates artists and appreciators of art, and it instructs in recreational activity.

Education, then, may be thought of as an accelerator, an activity which affects the rate of development of each function necessary for continuous progress toward the ideal of mankind: the ability of every individual to

---

[14] H. M. Groves, "National Economy and Education," in *Encyclopedia of Educational Research*, C. W. Harris, ed. (New York: The Macmillan Co., 1960), p. 921.

satisfy each of his expanding sets of desires. It may also be thought of as a decelerator of retrogression. A certain amount of education is required to keep us where we are. The amount of education required for this purpose is probably proportional to the degree of complexity and specialization in our way of life; that is, to the extent of our society's development. Therefore, in order to measure the complete value of educational output, it would be necessary to measure (a) the value of each social institution, (on some one scale), and (b) the rate at which education produces an increase or prevents a decrease in the value of the services rendered by these institutions. Clearly, we are not in a position to do this yet. We should, however, be devoting more effort to putting ourselves in a position to do so.

Problems of Measurement

What can we do in the meantime? There is a general feeling among those who have worked on the problem that at least the effect of education on economic development can be measured. For example, Groves observes:

> Search for a definite answer as to how much of available resources should be allocated to education goes on. No doubt the consensus of opinion among educators, including economists, is that the outlay is below the optimum. But the matter involves so many variables and value judgments that it is hardly likely to lend itself to a definitive answer. Allotment of resources to education, it seems, must always involve an element of faith. But the democratic decisions that determine this outlay can nevertheless be based on a great deal of evidence. Such evidence has been and is being mobilized by both economists and educators.[15]

Some effort has also been made by actuaries to evaluate the health benefits of education, but this has met with the same difficulties that we shall consider in connection with economic analyses.

In his very recent book, John Vaizey has reviewed the whole range of efforts to determine the rate of return to the community on educational investments.[16] A slightly earlier and briefer review can be found in Renshaw.[17] Although Vaizey finds some approaches preferable to others, all are open to many grave objections. This is not to say that these efforts are worthless; to the contrary, they are bringing us closer to a measure of the economic value of educational services.[18] Nevertheless, they do not

---

[15] Ibid., p. 917.

[16] John Vaizey, The Economics of Education (London: Faber and Faber, 1962).

[17] E. F. Renshaw, "Estimating the Returns to Education," The Review of Economics and Statistics, No. 42, 1960, pp. 318-24.

[18] See, for example, W. J. Platt, "Toward Strategies of Higher Education," staff paper, Stanford Research Institute, Menlo Park, California: January, 1961; V. L. Bowyer, "Relation of Public School Support to Subsequent Per Capita Wealth of States," Elementary School Journal, No. 33, 1933, pp. 333-45 and 417-26; and H. M. Groves, op. cit. (footnote 14).

yet provide a measure that we can use with any great amount of confidence for reasons which the economists themselves have been the first to point out.

Detailed discussion of the shortcomings of these efforts can be found in Vaizey's book cited in the paragraph above and in a supplement, edited by S. E. Harris, to The Review of Economics and Statistics.[19]

A few shortcomings of the efforts by economists to evaluate educational services are too important to skip over. In the so-called "cost-benefit" analyses of education, efforts are made to evaluate both the monetary input (cost) and economic output of the service. Although frequent comments are made in the literature on the exclusion of important non-economic outputs of education, reference seldom is made to the exclusion of important noneconomic inputs. A great deal more than money goes into the educational process: for example, the time, energy, comfort, and even safety of those involved. A complete evaluation of educational services should take these into account.

Most of the economic analyses of education consider investments only at the state or national levels, not at the metropolitan level, despite the fact that most of the investment in education is made by local authorities. Furthermore, most of the indices of economic well-being (e.g., Gross National Product or National Income) are obviously not applicable to urban areas without modification.

Since urban areas have relatively larger immigrations and emigrations than the nation and possibly the state, importation and exportation of educated people is an important consideration. The accumulation and utilization of "educational capital" is affected not only by investments in education, but by immigration and emigration, and these are affected in turn by many other characteristics of the community.

Because the urban area is much less self-sufficient economically than is a nation, I believe it will be much harder to develop adequate economic models of it. But it does seem to me that some important requirements of such models can be specified.

1. We must seek more than correlations between investment in education and economic well-being. We must seek explanations of the effect of education on economic well-being, and incorporate these into decision models of the type I have discussed earlier. That is, we must learn not only what effects education has, but how it brings them about.

2. With Platt[20] I agree that such a model will have to consider not only investment in education, but, in addition, at least investments in research and development, and in productive plant. Platt illustrates the kind of interrelationship that he suspects these variables have in the diagram shown in Figure 2.

It seems apparent that economic development depends on investments in research and development. The way investment in research and development affects the rate of economic development appears to depend on past

---

[19]Higher Education in the United States: The Economic Problems, Supplement 42, The Review of Economics and Statistics, 1960, especially "Summary of Discussion," by R. N. Cooper, and "Some Broad Issues," by S. E. Harris.

[20]W. J. Platt, "Educational Policy for Economic Growth," Menlo Park, California: Stanford Research Institute, August, 1961.

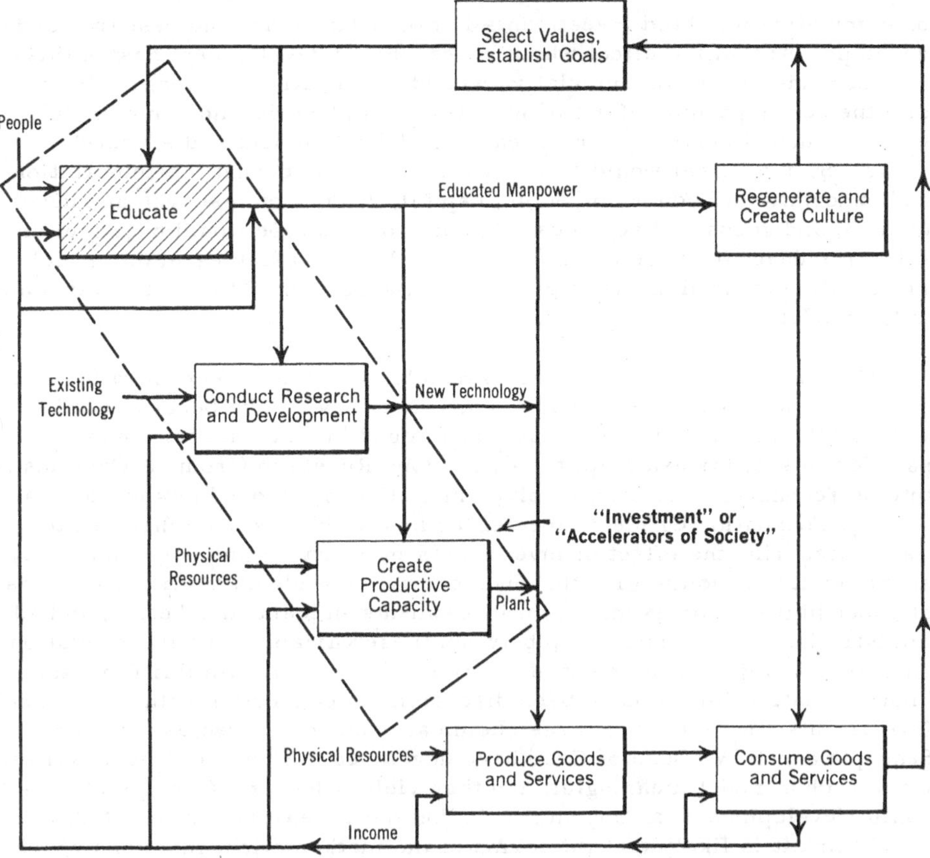

Figure 2. Education as a contributor to development. This diagram shows how education and other forms of investment contribute to development. The object of the system is "the good life," made up of cultural and economic attainment. The inputs to "educate" are people and education budgets. Controls on "educate" are values and goals of the society. Output from "educate" is educated people who perform the productive operations of the economic system, consume its goods and services, and create and regenerate its culture.

Source: William J. Platt, Toward Strategies of Education, SRI International Industrial Development Center Staff Paper (Menlo Park, California: Stanford Research Institute, 1961).

educational investments, because the return from investments in research and development depends on the education of the people involved in it. Similarly, the way investments in productive plant affects the rate of economic development seems to depend on past investments in both research and development and education. Consequently, the effect of investment in education on economic development seems to be largely indirect through its impact on research and development and plant-investment policies.

The current rate of increase in economic development depends on past investments in education. It will be more sensitive to what has been invested in some years than to what has been invested in others. Consequently, the variable describing the economic input to educational services will have to be a weighted accumulation of past investments.

It would seem, then, that an adequate decision model would state that the rate of increase in economic well-being in one year depends directly on cumulative weighted investments in production plant and research and development; that the effect which research and development has on this rate depends in turn on cumulative weighted educational investments; and that the corresponding effect of investment in plant depends on cumulative weighted investments in both research and development and education.

3. Such a model would also have to take into account the importation and exportation of educated people, capital, results of research and development, and goods and services. A model of this type would be useful only if the residual effects (those not explained by the variables in the model) are small compared to the effects which are accounted for by the variables in the model.

Can the relationships between variables which are necessary for constructing such a model be found? I am sure they can be because similar and equally difficult functions have been found in other areas. Over the past few years, for example, the Operations Research Group at Case Institute of Technology has been involved in a study sponsored by the National Science Foundation Office of Special Studies which has been directed toward measuring the effect of investments in research and development by a firm on its economic growth. Such growth depends on a variety of types of expenditures: for example, on investments in plant, marketing, and administration. The relationship between these variables and their relationship to economic development of the firm is not too dissimilar from the kind of relationship we have been discussing in connection with educational investments. In a study of three chemical companies, Professors S. S. Sengupta and B. V. Dean of Case have developed the functions necessary to make such a model meaningful, and they yielded better estimates of economic development than anyone, including the researchers, expected.

I think it is likely that the forms of the mathematical functions which are suitable for explaining the effects of educational investments in different urban areas will be similar; but that the parameters of these functions will differ. The similarity of form would reflect similar functioning of education in different communities. The difference in parameters would reflect the uniqueness of each community. This was found to be the case in deriving the research-and-development functions for different chemical companies.

Construction of the kind of economic model that I have been discussing is a formidable task, but I feel confident that it could be successful. It should be remembered, however, that at best it would provide only an economic evaluation of education. The value of the noneconomic inputs and outputs would remain to be dealt with.

The Need for a Broad Approach

At this point those who are skeptical about the capability of research to contribute to planning are likely to nod their heads knowingly and say, "See what I mean? In order to solve a problem which is already very complicated, you now find that you must solve one that is even more complicated and difficult. Do you call this progress?"

This reaction is based on the false assumption that in science questions can be arranged in a hierarchy of complexity and difficulty. Simplicity is not a characteristic of a problem but of our current ability to cope with it. Furthermore, science does not proceed from simple facts to complex theories. First, any fact, when probed in depth in circumstances where it makes an important difference, becomes as complicated as any theory and "in fact" requires confirmation of a large number of theoretical assumptions in order to confirm it. Secondly, facts are just as likely to follow from theory as theories are from facts. More specifically, we have come to realize that there are systems (and the community and science are among them) knowledge of whose parts cannot be developed continuously without simultaneously developing understanding of the larger systems that contain them.

The ability to evaluate any one service ultimately depends on our ability to evaluate every other service. Hence, progress must take place along a front, not by a series of deep but narrow penetrations into ignorance.

We restricted ourselves earlier to economic considerations in the hope that it would yield a currently useful, even if partial, measure of the value of educational services. Up to now it has obviously failed to do so. Would it be possible, then, to use an approach more like that used in dealing with transportation services? At first glance it may not seem so. Most of those who are receiving the service do not know what value it has or will have to them.

But what about those who are responsible for those who are being educated? Since they are responsible, are not the values which they place on education of their children or wards relevant?

Suppose we determine for the adults (or a sample of them) who are responsible for children who attend publicly supported schools, what is the largest direct charge per student per year that they would be willing to pay before withdrawing their children from public school, if school attendance were not compulsory. This does not seem to be a formidable task. The sum of the values obtained in this way would be an estimate of the <u>gross</u> value of the output of public education. It would be an underestimate because members of the community without children in public school are not included. But for them the appropriate question would be about the maximum amount of taxes they are willing to pay directly or indirectly to support public education.

In order to obtain an estimate of the net value of educational services, we would have to obtain an estimate of the value of the inputs. But following the outline that I gave in the case of transportation services, it does not seem impossible to get the parents' or guardians' evaluation of the students' time, energy, safety, and loss of pleasure (if any); and to determine the direct costs of education to them. The costs of some of these inputs, such as that for students' time, may be negative because the parents find it of value to have their children "off their hands" for part of the day.

I am not suggesting that such an approach to evaluating educational services is the best possible one, but it may be the best practical one. It would yield information which should be useful for political, as well as evaluative purposes. The principal objection that can be raised against it is that the judgments obtained may not be based on a knowledge of the consequences of education. But it should be noted that we don't have this knowledge either.

Finally, I should like to observe that the errors introduced by subjectivity and lack of knowledge may not be nearly as great as those resulting from exclusion of noneconomic considerations in building economic models. Where we do not know how to measure the total net value of an urban service, what right have we to impose on those affected by the service either our subjective evaluations or our incomplete objective evaluations and the conclusions derived from them? We have such a right only if we can demonstrate that our judgments serve the users of these services better than do their own; and this would be at least as difficult to demonstrate as it is to develop objectively based measures of the total net value of the service.

CONCLUSION

I have tried to show how quantitative measures of the value of urban services might be constructed. My hope is that the methodology I have outlined will be helpful and that my two efforts to develop measures will be at least suggestive. I do think some general conclusions can be drawn from these two efforts:

1. Using either objective behavior or subjective judgments, estimates of the values that individuals place on inputs and outputs are obtainable in principle, and are not prohibitively difficult to obtain in practice.

2. These evaluations can be obtained on a single scale such as the monetary one.

3. The classification of input objectives (involving cost, time, energy, safety, and enjoyment) is probably adequate to deal with any urban service.

4. Building up a measure of a community's evaluation of a service from the evaluations of individuals is more likely to yield useful measures quickly than is the effort to measure the total value directly, We should, however, push in both directions at the same time.

The kind of thinking that has characterized most of past planning can be called <u>synthetic</u>: involved with putting things together. Here I have tried to show how analysis--taking things apart--can play an equally important role in the planning process. Synthesis and analysis are the two legs on which decision making moves. We cannot make much progress no matter how much we extend one leg if the other is firmly anchored in place.

Furthermore, planning has been predominantly dependent on experience rather than experiment. Social experimentation must become a common tool of the planner. To be sure, this requires acceptance by the public that will be affected by the required manipulation. But, as I have tried to show, I do not think any noticeable short-run losses to the public are necessary. There will certainly be long-run gains. The major obstacle to the effective use of experimentation, it seems to me, is not the public's attitude, but the planner's. I see little evidence of his trying hard to experiment.

If you will indulge a selfish moment, I would like to express the belief that if urban planners and my fellow professionals in operations research were to collaborate, a more appropriate balance of analysis and synthesis, and of experience and experiment would be obtained than is at present. Such collaboration between national planners and operations research workers has proved itself in many countries. I am convinced that if it were to be turned on to urban problems it would be equally fruitful.

Finally, there is one important point which I should like to stress about evaluation of urban services. If my analysis of the steps involved in obtaining an evaluation of such services is correct, these steps cannot be avoided. They may be taken unconsciously and qualitatively, but they cannot be side-stepped. Until they are raised to consciousness and made explicit, they cannot be studied and systematically improved. Therefore, it seems to me, that efforts such as I have made here are justified if their only effect is to make planners expose to themselves and others their evaluative judgments and the procedures by which they are reached. Awareness of one's own limitations is the foundation of all progress.

. . . . .

I began work on this paper with a great deal more confidence than I have as I submit it for publication. The more deeply I have probed into the problem of evaluating urban services, the more difficulties have been revealed. Most of these difficulties are ones with which experienced planners and economists have long been familiar. I have not managed to put any planning problems to rest; if anything, I have stirred them up a bit. Perhaps this will attract more research to the planning process.

The methodological problems which I discuss here are not unique to my approach to evaluating urban services; they are inherent in the problem and, hence, are present in any approach to it. Too frequently, however, they are not treated explicitly. Although frequently hidden, they are never avoided.

This is an optimistic paper, but I feel some optimism is needed here. In my experience in working on a wide variety of large systems, many of the difficulties which appear in principle disappear in practice. Feedback from work "in the field" almost always seems to reduce methodological difficulties. Unfortunately, I have had to write this paper without the benefit of such feedback, and it shows. Therefore, I hope the reader will treat this as a working paper. Much more work on the problems discussed in this paper and the paper itself is needed.

6

Benefit-Cost Analysis as a Tool in Urban Government Decision Making

by Nathaniel Lichfield and Julius Margolis

Benefit-cost analysis is the name commonly applied to that branch of economic analysis which is used when making decisions about public investments. While it has wider applications, and can be used in the private sector, it is most useful in the public, where it applies the conceptual framework of economic theory, both normative and positive, to problems in which the market competitive forces and private incentives for gain do not provide straightforward guidelines for choice. In the public sector the penalties and rewards usually considered in economic analysis are absent and the prices so central to quantitative analysis in the private economy do not bear the same meaning. Revenues are not necessarily benefits and expenditures not necessarily costs, and their differences do not necessarily measure a goal to be sought. The substitution of benefits for revenues and costs for expenditures necessitates going beyond the traditional confines of the economist, but not outside the economist's model of choice. Success in the venture is most likely in those public activities which most closely approximate private economic activity. This is attested by the initial development of the tool in water resources[1] and highways,[2] where services are produced for actual or notional sale. But we believe that as we gain experience in measuring the imponderables and in understanding the political and administrative processes of government, the techniques may become applicable to a fuller set of public activities.

We are not primarily concerned with the decision making process as such in urban government, but rather with showing how a particular

[1]For example, Roland N. McKean, Efficiency in Government through Systems Analysis (New York: John Wiley, 1958); Otto Eckstein, Water Resource Development (Cambridge: Harvard University Press, 1951); A. Maass, M. M. Hufschmidt, R. Dorfman, H. A. Thomas, Jr., S. A. Marglin, and G. M. Fair, Design of Water-Resource Systems (Cambridge: Harvard University Press, 1962).

[2]For example, Herbert Mohring, The Nature and Measurement of Highway Benefits: An Analytical Framework, A Transportation Center Research Report (Evanston, Ill., 1960); Tillo E. Kuhn, Public Enterprise Economics and Transport Problems (Berkeley: University of California Press, 1962).

analytical tool, benefit-cost analysis, can and should be used to improve the process. To embark on this we must clearly have in mind a model of the decision making process, for any terminology can have precise meaning only in relation to a particular model. This raises difficulties, for urban governments do not have standardized processes; a particular government may employ many, and more than one normative model has been put forward.[3]

ELEMENTS OF DECISION MAKING AND BENEFIT-COST ANALYSIS

The elements of the decision making model, within which benefit-cost analysis is used, are few but difficult to formulate for quantitative analysis. A decision is a choice among feasible alternatives. The feasibility condition arises because of technical, legal, and organizational constraints upon choice or alternatives. Assuming that the behavior is purposive the choice involves a valuation of the consequences of the alternatives and a criterion to choose among them.[4] In the evaluation, goals should be formulated and then, usually, translated from subjective values to objective indices, i.e., from psychic goals of welfare maximization to instrumental objectives such as housing standards, school facilities, traffic flows, etc.

The decision making model can be used systematically for goal formulation, constraint identification, target specification, criteria, and final benefit-cost analysis of alternatives, but this sequencing is unnecessary. In fact, we shall suggest that it will not be followed in the strategy of decision making for urban government.

The range of possible decisions to which analysis can be applied is wide. It will be most successfully applied to those cases where we have repetitive events, simple goals, and well-defined products. Repetitive events will permit us to forecast with some confidence the consequences of action; simple goals enable us to apply value weights which will be found acceptable by the decision makers; and measurement will enable us to balance the consequences. But these are just the cases where experience unguided by analysis is most helpful in shaping decisions which are tolerably sensible. Paradoxically, analysis in the areas where information has been most meager and the analyst speaks with least confidence may have the highest payoff.

The formulation of goals is the most difficult and perplexing part of the analysis. In consumer economics the analysis of goals is sidestepped by the simple assumption that the household seeks to maximize utility without identifying any of its dimensions. In the analysis of the firm, the dominant tradition is to choose a single quantifiable goal, profits, as the desideratum. At a very general level it may be correct to say that the individual seeks to maximize utility, the entrepreneurs seek to maximize their utility, and, by analogy, the body politic should seek to maximize an

---

[3]Herbert A. Simon, Administrative Behavior, 2nd ed. (New York: Macmillan, 1959); Martin Meyerson and Edward C. Banfield, Politics, Planning and the Public Interest (Glencoe, Ill.: The Free Press, 1955).

[4]Where the behavior is not purposive the decision might be instinctive or unconscious, but it can be explained by reference to a purposive model. The particular choice will imply a particular goal, and its preference over another, etc. But the choice will not necessarily be "rational" in the sense that it represents the most efficient choice.

index of social welfare. But while utility or welfare maximization may be a useful heuristic device for the analyst concerned with explaining behavior, it has little value for the normative branch of science which seeks to advise the individual, the firm or the government on how to decide. The analyst must be able to identify the dimensions of utility or welfare, i.e., what are the desired qualities of goods or services. In the case of households it would mean such dimensions as security, shelter, nourishment, recreation; for the firm it could mean income, stability, certainty, power, quiet life, etc. Identifying the dimensions of welfare is a necessary step in making goals operational.

Though we may not be able at this point in time to specify an index of national security, or optimal health, or democratic participation in government, or other similar goals which are widely held by the American public, we can identify lower-level goals.[5] For instance we may not be able to identify the optimum pattern of shelter--the free standing house or apartment or row--or whether this dimension of shelter is the most meaningful. But we do know that there is a community consensus about such things as the internal bathroom, sanitary facilities, lighting, and floor space per capita. These dimensions of a satisfactory dwelling unit become the basis of the definition of unsatisfactory housing. The definition can be readily criticized, but despite its shortcomings the set of quantitative attributes of a satisfactory dwelling unit is more useful than vague descriptions of insanitary or inadequate housing, and certainly it is more susceptible to analysis and improvement.[6]

The above process of settling on goals short of maximum community welfare has been referred to as sub-optimization.[7] As the label clearly implies we are not doing the best that could be done. It is unpleasant to recognize that our choices are most likely not the best, but the forces of ignorance, administrative complexity, and political pressures make utopian any insistence on nothing but the best. Our recognition of the sources of less-than-optimal choice can help us overcome some of its shortcomings and enable us to do better.

A willingness to accept operational surrogates for our goals does not settle our problems but it does make them more manageable. How are these surrogates found? How can they be improved? Are they additive since they do not have the dimension of welfare? Though we shall discuss these questions throughout the paper a few comments at this point would be in order.

Most decisions are taken to improve an existing situation. This trite statement is obvious to government officials but unfortunately it has not found its place in the normative models of public services. It is the dissatisfaction with existing conditions which can be the basis for the exploration of desirable objectives. The dissatisfaction can find expression in a formal review of the operation of services, as it did when the City of

---

[5]Some definition of national goals is in Goals for Americans, the Report of the President's Commission on National Goals (New York: Prentice-Hall, Inc., 1961).

[6]American Public Health Association, An Appraisal Method for Measuring the Quality of Housing, 3 vols. (New York, 1945, 1946, 1950).

[7]Charles Hitch, "Suboptimization in Operations Problems," Journal of Operations Research Society of America, Vol. I, No. 3, 1953.

New York Board of Estimate in 1950 set up a special committee to conduct a comprehensive survey of all city departments and agencies "in an effort to achieve the greatest amount of efficiency and economy possible, consistent with the maintenance of adequate public services..."[8] Or it might result in the ad hoc investigation of a particular service by, for example, a work-measurement study. Of course there is no guarantee that a metric can always be assigned to the desired product but we feel confident that a probing analysis of assumed benefits can clarify objectives and thereby make more precise the later assessment of benefits. For example, a post office with long queues is unsatisfactory. An expansion of counter space is desirable. The survey prior to the determination of the new counters could and should contain some estimates of reduced waiting time associated with the additional counters. Of course, not every problem has transparent objectives. For example, there is no consensus about whether we have low educational levels in our schools. To some degree this is because education is a composite good and there is no agreement about the desired product mix, i.e., citizenship, sociability, judgment, knowledge, etc. Even here where some of the goals are highly "intangible" and there may not be a satisfactory consensus about the goals we may be able to identify desirable actions on the production side--better trained teachers, a lowered pupil-teacher ratio, etc.[9]

The identification of reasonable measures of government product, creative and useful as it may be, is still a far cry from the solution of our problem. Values must be assigned to the products. In some cases we have made considerable progress, for example, transportation, fire, sanitation; for other services--education for instance--we have a beginning; while for some important services, such as police and social services, very little analysis has been done. In all cases the test of value weights is found in the answer to the question of what the users are willing to pay, either directly as consumers, or through their government as representatives. In some cases we can proceed with a model of a simulated market and ask about the demand if the public service were priced in a free market. This has been the avenue taken by transportation studies. Their major index of demand has been cost-savings in transportation. But for other services the dominance of social policy considerations will make this useful model too restrictive. There are ways of analyzing social policy goals quantitatively. We may be able to infer from the behavior of governments the implicit weights they have assigned to the public services. More usefully we could confront the legislative and top administrative decision makers with the costs of procuring incremental amounts of the different services and from the resulting decisions might begin to construct a matrix of value weights.[10]

Coming now to constraints upon choice, it is clear that greater precision contributes to better analysis in the same way as greater precision in goal formulation. In fact, the constraints result in closer definition of

[8] Modern Management of the City of New York, Report of the Mayor's Committee on Management Survey, Vol. I (1953), p. v.

[9] John Vaizey, The Economics of Education (London: Faber and Faber, 1962).

[10] Maynard M. Hufschmidt, John Krutilla, and Julius Margolis, Report of Panel of Consultants to the Bureau of the Budget, Standards and Criteria for Formulating and Evaluating Federal Water Resource Developments (Washington: U.S. Bureau of the Budget, 1961), mimeo.

goals--so much so that it is difficult to distinguish between constraints and the instrumental objectives of goals. The goal of a house with constraints as to location may be no different in practice from a goal of a house in particular locations; and a series of constraints may imply a particular goal. But in concept they should be distinguished. The constraints limit choice and the instrumental objectives define the goal. In particular situations the distinction may become of significance.

The interplay between constraints and goal is especially significant in the case of public services where the bureaucracy is legislatively and practically bound to formulate its programs within the confines of existing legislation and administrative structure. But the constraints of organization and legislation are not sacrosanct. If the analysis should reveal that a large payoff would result from the merger of the police and fire departments, the constraints of independent departments should be subjected to analysis. Or, if a constraint of zero price shows that the costs of service at the margin are greater than the benefits, a rationing system might be introduced. It might be utopian to expect that all institutional and legislative constraints could be abandoned in the analysis, but an essential element of suboptimization is the constant search of methods to remove the factors causing less than optimal behavior.

Coming to the criterion, there is no all-purpose formula, and several possibilities are available. Examples are the maximum difference between benefits and costs; the maximum benefit for a specific cost, or for any cost below a given maximum; the minimum cost for a specific benefit, or for any benefit above a given minimum. But these criteria cannot be used at random. The criterion chosen will be affected by the alternatives being considered, the particular constraints, and kinds of measurement being employed.[11]

Finally, as to the choice. This, in effect, is the proposal to buy a measure of a goal at a particular cost. The clarification of goals and constraints will have helped in assessing the goal; and the clarification of constraints and criteria, in assessing the relation between goal and cost. But the decision is still to be made, and at this stage of social science the benefit-cost analysis will not necessarily give us a clear answer as to the best choice. First, most decisions will achieve goals in varying proportions, and since the welfare contributions of all services are not comparable the decision makers are in the position of implicit valuers. They must decide which of several packages is preferable and whether a set of costs are justifiable by a set of benefits. As such they are in no more difficult a position than the individual consumer who buys one of a set or bundle of goods without having an index of how each good adds to his welfare. But their position is much more difficult in another sense since "social choice" is involved, for the decision is being made by one person or body on behalf of others. The decision maker must balance off the costs and benefits which accrue to different individuals, a complex issue in which unraveling in recent years has not been absent.[12] A third element of judgment which enters on a lower level deals with the accepted facts. The future can only

---

[11]McKean, op. cit. (footnote 1), pp. 34-49.

[12]Jerome Rothenberg, The Measurement of Social Welfare (New York: Prentice-Hall, Inc., 1961); and C. W. Churchman, Introduction to Operations Research (New York: John Wiley, 1957), chaps. 4-5.

be known with varying degrees of uncertainty and the decision maker must make a judgment of the alternative possibilities.[13]

SOME SPECIAL ASPECTS OF URBAN GOVERNMENT

Urban government is not free to adopt whatever functions, and therefore whatever goals, it likes, for it is subject to a bewildering array of influences. The municipality is a "creation of the state," a fact which grows in significance when the traditionally rural bias of most state legislatures is taken into account. The state's constraints are felt in a variety of ways: in legislation which grants specific powers; in the degree of grant of home rule; in the extent of delegation of law making power; in the kind of state administrative supervision and control.

The state's power in circumscribing urban goals is not, however, absolute, for it must itself act within its own and the federal constitution. This brings in the courts as an active party in the cities' definition of function and goals.

But within these constraints, the choice of function and goal lies primarily with the policy makers of the urban government. Just who these are in any particular city varies with the form of government: weak mayor--strong council, strong mayor--weak council, commission or council-manager. But whoever they are, they are not the sole goal formulators, for they share the function with five other influences. First, the electorate, or rather that part of the electorate whom the representatives set out to represent. The successful candidate must have some regard for his platform, however minimal. There is not only the next election to consider, but also, in some states, the means of popular control between elections, by initiative, referendum, or recall. Second, there is popular pressure: the radio and press, acting as the "voice of the people"; citizens' organizations which act as watchdog and often the spearhead of new policies; citizens' boards which are nominated to advise on and sometimes direct certain functions--parks, libraries, city planning. Third, the "power structure," the industrial, commercial, and financial interests which "set the line of policy" for the community and are able to see it carried through.[14] Fourth, the "invisible government" of Elihu Root, the party machine and party bosses who so often control the elected representatives, either to good or bad. Finally, the city administration. Their division of responsibility with the policy makers, and the commissions these set up, varies considerably. But although the administration cannot formally exercise discretionary power, but only delegated power which is ministerial or administrative, it nonetheless does more than merely carry out policies; it is instrumental in determining what those policies shall be.

In this situation of so many competing influences on policy, the definition of particular goals, and the resolution between goals, falls within the arena of "politics"--the activity by which conflicting ends are resolved

---

[13]G. L. S. Shackle, Decision, Order and Time in Human Affairs (Cambridge, England: The University Press, 1961).

[14]Floyd Hunter, Community Power Structure: A Study of Decision Makers (Chapel Hill: University of North Carolina Press, 1953).

to provide a basis for a decision.[15] In so far as the attempted resolution is objective, it would attempt a social evaluation of goals, that is, a definition of the public interest. This, however, is notoriously elusive of definition. Who are the public in an urban area: the residents; the people who work there, including commuters but excluding those who travel out; the people responsible for the economic base of the community? And, once this has been defined, is the public interest the sum of individual interests or something separate? And in either case, how is it ascertained?[16]

The institutions of urban government thus make the definition of its goals a very complex matter, and they also impose constraints upon local choice. Such constraints must vary considerably with the nature of the function being exercised, but only four constraints of a general nature are discussed here. First, all urban governments have a defined territory of jurisdiction which, once determined, is changed only with difficulty. Since such boundaries cannot be suitable for all the different urban functions, and become out of date for functions they once suited, an urban government often has territorial constraints upon its quest for efficiency: in the unsuitability of its area and in the activities of neighboring authorities carrying out similar functions. Consequences are, of course, most acute in metropolitan areas. The remedy, metropolitan government is still a long way off but some relief is currently obtainable: through cities having extraterritorial powers, establishment of special districts, voluntary municipal co-operation and advisory regional planning. But the relief is small, so that decisions resulting in optimum efficiency for services are not easily attainable. Second, municipalities have budgetary constraints in the form of varying fiscal capacities. Therefore, even if preferences for public services were the same, the levels will be quite different. But a social judgment in favor of uniform public service will induce higher level governments to use grants-in-aid or direct regulation to encourage a specific level of service, or at least to make the local costs of carrying on some services quite different from the local costs of others. The result is a varying level of standards, a level which state and federal governments try to raise to at least minimum levels when operating a grants-in-aid program. Third, local governments operate under capital and revenue budget constraints, because of the state limits on the maximum rates and debts and restrictions on types of taxes they can use. This will aggravate the problem of capacity variability but the implications for analysis and policy are quite different. Here there is a problem of reconciling differences within the community. Fourth, some services have the constraint of functional, as opposed to fiscal, interdependence. Instances are highways and mass transportation, urban freeways and downtown car parks, industrial zoning and public housing, social welfare and urban renewal. Decisions on the one would affect decisions on the other, whether the related decisions are made by one municipality, special districts or authorities, or higher level governments.

---

[15]Meyerson and Banfield, op. cit. (footnote 3), pp. 304 ff.

[16]Glendon Schubert, The Public Interest (Glencoe, Ill.: The Free Press, 1960).

LOGIC OF BENEFIT-COST ANALYSIS IN URBAN GOVERNMENT

In formal terms, the model of analysis we have in mind is as follows: Each department is to specify its production function, spelling out its assumed goals, constraints, and criteria. The production function is to contain alternative ways to achieve the agency's objectives and to contain its evaluation of the consequences and the payoffs in extending its service along different lines. The various departments would submit their programs to the policy makers. These would review and return the programs to the departments for reformulation within an amended framework, which would incorporate the definitive goals, constraints, and criteria. The working of this basic model would be assisted if either or both of the following courses were followed: The departments would work with a co-ordinating committee or chief co-ordinator; or some guidance could be given by a chief executive after consultation with the policy makers.

The above sketch of task assignments may be rational, but governments, like other complex institutions, are not fond of the disruptive consequences of rational decision making. Members of the bureaucracy know their roles and find convenient the traditional ways of doing things; they do not feel the need to define goals, constraints and criteria or alternative courses of action. But let us not establish too firmly the image of a tradition-bound government. Bureaucratic inertia exists, but it is subject to shocks which force the government to review and be deliberative about its actions. Traditions have been the basis of most of the criteria for public choice, and such traditionalism is easy to understand, but there are forces which can and will cause traditional criteria to be abandoned.

The problems of urban areas have been repeatedly enumerated. Their growth continues unabated. Industry and population spill over into the countryside while the build-up in the center continues. The build-up of the old is less dramatic, but it still remains the heart of the city. Conflicts in development are exacerbated with growth. Governments are called upon to mediate: to plan and to provide services to facilitate the growth. A government can remain obdurate and seek simply to attempt to duplicate existing patterns. But this is doomed to failure. The changing structure of the city will not tolerate the old. Pipes, pumps and water pressures will change to fit the width and height of the city. The police will motorize to control the motorized public. How shall these and many other changes in services and controls be effected? To what purposes shall the additional government resources be devoted? The logic of benefit-cost criteria suggest a strategy of decision making analysis.

An ideal form of decision making on these problems would have the administration prepare reports on all of the alternatives facing the policy makers, identifying the costs and consequences of each alternative, following some definition of goals, constraints and criteria. The policy makers, knowing the dimensions of welfare of its urban constituency, would choose among the alternatives so as to maximize the welfare of the city's present and future population. No vested interest could claim inviolability. The politicians would know and serve the welfare of the community; the administrators would assiduously search out and evaluate all relevant alternatives. But, we hasten to recognize, the idealized form is too far removed from reality to be a feasible model. The councilmen do not identify the goals of government, and therefore they cannot quantify the dimensions of welfare, even if it were true that they accepted the obligation of serving the public interest. The administrators will neither search out all alternatives, specify all consequences, or await on the wisdom of the council as to the evaluation of consequences.

But the realities of urban government decision making do not imply the abandonment of the benefit-cost analysis. As we argued above, benefit-cost analysis still has a place even if decisions are taken on impulse or as a result of bargaining: as a post hoc test of the decision, or demonstration of its rationale, or basis for testing its results. Nonetheless, most profit will come from the use of the analysis if it is part of some system of decision making, similar to but not necessarily identical with a "rational" model. The question therefore arises, whether there is a model of decision making which is a reasonable approximation of current urban government practice and which allows for a profitable use of benefit-cost analysis. Possibly a more radical question is more appropriate. If benefit-cost criteria are sensible tools, can the structure of decision making in urban governments be sufficiently altered so as to permit the better use of the criteria? We stress the latter since the development of benefit-cost analysis is more than an analytical tool sought for by public administrators. Its use is dependent upon a more rationalized structure of public choice, potentially of enormous value in helping governments, and not only urban governments, face up to many of their most pressing problems.

Let us abandon for the moment the long term hope of a welfare-maximizing government. Instead, let us adopt the more currently realistic model of a deliberative government. This government continuously seeks improvements in its current operations, as well as the best use of any additional funds, instead of uncritical expansion of existing programs. We suggest that every agency, prior to a request for new funds, should provide an analysis of the current operations of its programs. While such an analysis would be more in the nature of a work-study than the kind of investment forecast we have been discussing, it could be made in terms of benefits produced and costs involved, as we have used these terms.

From such results the agency should be able to take four constructive steps. First, it can prepare indices of expenditure per unit of service or product (unit costs). These it can use for comparison between services. Second, it could use the indices of expenditure as a basis of preparation of performance budgets. Third, in the revenue budget it could consider reallocation of variable costs among particular related services with an eye to payoffs in benefits. Fourth, in the capital budget, it could compare the benefits to be obtained from the marginal transfers of investment funds, within the departmental budget constraint, to the various services. The comparison would be made among related services and also extended to the less similar activities of a department. For this, as with the revenue budget, it is desirable to translate as far as possible the outcomes of the various operations into comparable performance units. A health agency should be able to compare an expansion of hospital facilities with outpatient facilities and both with health protection measures. A police department should be able to compare increased motorized patrols with foot patrols. Finally, in a cross-departmental analysis, the highway department should be able to compare its cost of saving of life by auto accident prevention measures with the fire department's cost for saving life by fire prevention.

Identifying Goals, Constraints, and Criteria

Such self-conscious study by departments of a deliberative government requires the identification of goals, constraints, and criteria. How is this to be done?

Goals and constraints can be considered together. On these, we would reiterate that the absence of goal formulation does not mean the absence of a goal. This is built unconsciously, however, into any decision, and is achieved by any action to implement that decision. The corollary is that if the goal is not formulated by the policy makers it will be by the administrators. Which course is preferable?

We think that the advantage lies with the policy makers. First, it is their responsibility and not the executives' to express the values which should be reflected in goals; the responsibility cannot be avoided for they will express the values anyway in making the decisions. Second, however vague they are, they give a certain direction to the executive and so to some degree, and often to a large degree, limit the alternatives which the executive need explore. This clearly is necessary in relation to constraints; where the proposals of a particular department has implications for the operations and budgets of other departments, the high-level decision is first needed as a guide to sub-optimization.

But it must be accepted that within the context of partisan politics, policy makers will express their real goals only in the vaguest of terms, and will adopt explicit goals which they cannot, or do not intend, to follow through. Their reasons are not necessarily base. They may know that while there is a widespread consensus on many goals, these would not necessarily attract open universal support; sectional interests might fear the loss of bargaining positions on other goals which are not universally supported. An instance was the reluctance of the British Trades Union movement to join in central government economic planning, which they support, because of its implications for surrender of rights to wage increases unrelated to increases in growth.

To expect politicians to abandon this kind of political reasoning in the foreseeable future is expecting too much. This does not however imply abandoning the framework for the deliberative analysis. For one thing, since all decisions imply goals, an analysis of past decisions can provide a schedule of concrete, actual goals as a basis for future choice. Such an analysis would reveal precedents in the level of educational services. Unless the school board, expressly or by implication, wishes to depart from its current standard, then the goal in this respect is clearly formulated and no other need be followed in preparing proposals for new schools. This procedure would still leave room, of course, for a questioning of these standards, either by the executive or board itself. And where certain high level decisions have been taken--for example, to redevelop the downtown central business district--then implicit or explicit goals in relation to suburban shopping centers (in the same city) are clarified in consequence.

This analysis of past decisions does not necessarily, of course, give the right guide to the future. For this, in the absence of a lead from the policy makers, there is considerable sense in the development of goals, constraints and criteria at the operating rather than top level. The departments could individually specify the framework, on lines on which they would hope for endorsement. They would then put up their proposals, direct to the policy makers, optimizing within their self-imposed framework for the particular department. The decision makers would then, before making any decision, have to consider the possible disharmonies in the individual frameworks: total budget constraint, repercussion effects, inconsistent evaluations of benefits, conflicts in goals. They would then resolve the inconsistencies, and remit the proposals to the departments for review.

This procedure is clearly likely to be uneconomical in effort. It would be improved if the heads of the different departments were to act in accord on setting the framework, so ironing out any avoidable inconsistencies, or were to be co-ordinated in their activities by the chief executive. Economies of this kind have been realized in those cities which have appointed urban renewal co-ordinators.

There is a positive advantage in this procedure over attempts at prior precise goal formulation: value weights are discovered via behavior rather than attempted ab initio. It is thus in the operations of an ongoing program that implications for welfare of choices can be seen and it is here that operational analysis can derive the opportunity costs of alternatives, and force the council to participate in the hard act of choice by attaching explicit weights to outcomes.

We introduced this section by stating that we held little hope for an optimal model of decision making. The structure we suggest would seem much more easy to achieve. Possibly it never will be fully adopted, but we believe that it will prove to be a feasible framework for a decision making model which is consistent with the use of benefit-cost analysis.

SOME METHODOLOGICAL COMMENTS

Social Evaluation of Benefits

The general approach to benefits evaluation has been to define benefits as the payments the consumers of the service would be prepared to pay. This is a fairly adequate formulation so long as we are discussing products like water which are almost identical to the usual run of commercial products, but when we turn to most of the services provided by urban governments we are confronted by opaque questions of public policy. For many public services the individual's evaluation is dependent upon the quantity received by others, as in education; in some cases (police control) an individual will tolerate the "service" only because it is applied to others. If the analysis is to be useful beyond a few commercial type services it must find politically acceptable ways to cope with the problem of political evaluation of social goods.

For commercial type goods, two bases of individual evaluations are used: (1) cost reduction and (2) productivity or satisfaction enhancement. A minimum value to a road improvement could be estimated at the value of reduction in travel costs. The power of this approach is seen in Marion Clawson's method of assigning benefits to parks.[17] Essentially he converts a problem of satisfaction enhancement into one of cost reduction, by

[17]Marion Clawson, Methods of Measuring the Demand for and Value of Outdoor Recreation, Reprint No. 10 (Washington: Resources for the Future, 1959).

valuing the increased product of a recreation site in terms of the reduced costs to travel to a nearer site or to a site with added features. More directly, cost reduction methods enter by the analysis of the saving of private outlays, as in the relating of fire costs to reduction in insurance rates;[18] the relating of water softening to the reduction of laundering costs and preservations of heating equipment and clothes, etc.[19]

The measurement of productivity or satisfaction enhancement is more difficult. Where the product serves as a productive input we might estimate its marginal productivity and assign this as a value. This is done for irrigation water. It could be used to value that part of improved transportation which will lead not only to reduced costs of existing users but also to a generation of traffic based upon the increased productivity of the road system.[20] This approach can be extended to services which are usually perceived as final products but which can be analytically treated as intermediate products; education, for example, can be considered an input for labor skills. An estimate of the derived demand for a product is far simpler than an estimate of the direct demand. In the case of the derived demand, as education, we can reasonably assume that the scale of the purchasing industry is only barely affected by the increased supply and that prices of the final product as well as cost relations in the purchasing industry are only marginally affected, so that the derived demand can be computed.

For the direct demand of final product, satisfaction enhancement, we do not as yet have good guides. Certainly education has individual benefits which are not marketable via increased labor productivity. For some commercial type consumption services, as recreation, there may be a large enough private sector so that an approximation of a demand curve may be possible, but for others we may have to be satisfied with physical measurements. With these, some refinement is possible through subdividing a benefit into component qualities, awarding points to each by subjective evaluation, and then rating for quality. The additivity assumptions needed to permit one to add up such a tally sheet will rarely be realized, but still these exercises are of value.[21]

More often the problem of benefits measurement is more complex than the measurement of a non-observed demand function. There are few public services which have not become heavily endowed with overtones of public policy. Even as commercial an activity as water provision becomes a vehicle for achieving political objectives such as income redistribution, area development, etc. And even where these goals are not actively pursued an aura of public policy develops around the current operations of a public service so that it is often difficult to assert convincingly that the benefits are primarily those of the individuals who receive the services. Be it a valid case of public goals, an illusion of tradition, or a device to protect vested interest, benefits beyond those ascribable to the aggregation of the individual consumers must be assessed for most services.

---

[18] Geo. S. Nolte, Milpitas County [Cal.] Water District Study (1960), pp. 14-15.

[19] Engineering-Sciences Inc., Water Requirements for the City of Fremont [Cal.] (1960), pp. 58-59.

[20] Mohring, op. cit. (footnote 2).

[21] C. West Churchman, Prediction and Optimal Decision (New York: Prentice-Hall, 1961).

Generally the approach that must be adopted is that of consumer research. The decision maker, be it the council, manager, or administrators, is presented with a statement of the program's proposed accomplishments and costs. A decision to act contains an implicit evaluation of the benefits.[22] If the costs exceed the sum of individual benefits, an adoption of the program contains an implicit evaluation of the public policy benefits. One can compare the effectiveness of different programs in achieving the same public policy benefits. And if these are illusory the statement of costs and implied benefits might encourage the rationalization of a program.

The evaluation of public benefit by reference to benefit to a particular government is an approach which has become somewhat popular in the last decade. It is adopted, for example, in cost-revenue studies, which adopt as a criterion the differences between the costs and revenues of the government. The development of this criterion has not been sophisticated.[23] Generally it is used as a condition of adoption of a program, e.g., annex this area if the increment in taxes is greater than the increment in municipal costs. It is extended to cover the expansion of services: expand so long as the increase in revenue is greater than the increase in municipal costs. Since below we comment on a cost-revenue study, we shall make only one point here. Though a democratic government may represent the people, its fiscal status does not necessarily move in the same direction as the welfare of the community.

Though we are dependent upon the political process to judge the public policy benefits, scientific research can go beyond the simple presentation of facts. Many thousands of cities are spending funds today to accomplish these policy objectives. A study of their revealed preferences can provide some insight into the current political evaluation of these services. These figures are not prescriptions for the average or any city but the study of their determinants will provide valuable insight into the choice processes of government.

Though every municipal service has elements of both individual and political evaluation, we can classify the functions into five groups and characterize their benefits evaluations as follows:

1. Quasi-private. Where cities conduct enterprises that could readily be in the private sector--parking lots, golf courses, auditoriums--the benefits can be measured at market prices.

2. Self financing utilities--gas, water, etc. These differ from the last in that, because of the monopoly element, benefits usually cannot necessarily be valued at tariff prices, for these may not be market prices. Benefits should be valued at imputed prices, or measured in physical terms.

3. Engineering services--roads, sewers. Benefits are often widely dispersed. Sometimes they can be assigned as cost

[22] S. A. Marglin, in Maass, et al., op. cit. (footnote 1), pp. 67-87.

[23] William L. C. Wheaton, "Application of Cost-Revenue Studies to Fringe Areas," Journal of the American Institute of Planners, Vol. 25 (1959), p. 170.

reductions for users, and sometimes they are found in the increases in land values of the benefited property.

4. Social services--schools, health, fire-fighting, etc. These benefits have no market price, and while imputed prices can be estimated for some, for most they cannot be, as yet. Supplementing the studies of political evaluations there are several measurements which can appropriately be used. In their pioneer study, Ridley and Simon suggested (apart from financial cost) four types of measurement which would cover all municipal services.[24] These were in terms of effort (man hours); performance (physical quantity); result (the effect of effort or performance in relation to its objective); and service needs or problem magnitude (e.g., number of children at school age as a rough index of the need for education services). Any of these measures can then be related to their money cost to obtain criteria.

5. Police power. This includes a very wide variety of regulative function, typically to impose costs on particular people to the public benefit. The fact that cost and benefit fall on different parties, and by definition the "transaction" often falls outside the market process, need not preclude the use of a criterion. But the measurement of cost and benefit might be rather more complex than in the preceding groups. Where, for instance, the regulation restricts an established use (e.g., a new ordinance relating to smoke pollution), the estimate of cost to established property owners must reckon with both the cost of new works and the resultant value of the property, and the benefit to the public will be measurable by one of the four methods outlined above.

Discount Rate

Few topics in the benefit-cost analyses for federal projects have been more controversial than the appropriate discount rate.[25] But as yet in municipal planning the selection of a discount rate has not been a debated issue. Cities have generally accepted their borrowing rate as the appropriate discount factor by which to weight the benefits and costs of different years. But there is as little logic in the municipality accepting its borrowing rate as the discount factor as the federal government doing so. Abstracting from issues of uncertainty in dealing with specific alternatives, the municipal discount factor should, we think, be computed on the same basis as the federal.

A discount rate is a weight function for income of different years. This weight function is derived from a balancing of the forces of the utility of income distributed over the years and the productivity of savings (that part of income which is invested and therefore deferred to consumption to later years).

---

[24]Clarence E. Ridley and Herbert A. Simon, Measuring Municipal Activities, 2nd ed. (Chicago: The International City Managers Association, 1943).

[25]Eckstein, op. cit. (footnote 1), pp. 94-104.

The denial of consumption is in effect a cost to be balanced against the gains of increased income to be earned by productive investment. If the competitive capital markets were optimally organized to represent these social and private choices then the market rate of interest would be the price which could be used to measure the balancing of the forces of utility and productivity. But if one accepts the common view that the capital markets are highly imperfect, and further that an aggregation of individual choices even in a perfect market would not approximate the social consensus, then the market rate of interest does not provide proper guidance. For federal projects a reasonable proposal has been to solve for a desired growth rate and then the implied interest rate which could be used for project evaluation.[26] If this model is accepted for the federal discount rate then the same rate should be the price of capital made available to all savings and investing sectors of the economy, private or public.

The choice of this social discount rate is only part of the problem. If the federal government does not establish institutional forms, so that all private and public units will accept this weight function of future incomes, we can expect that each unit will use rates based upon their own access to credit, evaluation of risks, etc. The city with rapid growth and debt limits would have, implicitly or explicitly, a rate which is higher than a city with stationary economic prospects. The model of analysis developed for a national government could be applied to a city government assuming that local resources must be taxed to support local investment, and an appropriate discount rate derived. As a result a structure of rates would develop. The average of these rates need not be equal to the federal discount rate since the federal choice would be based upon an evaluation of national economic objectives which have quite a different fiscal and political base than the local communities.

Acceptance of the view that each local government solve for its own discount rate is consistent with the tone of sub-optimization raised in this paper. We assume that the metropolitan area should design its highway system, given the network of federal and state highways. It cannot design its system on the assumption that the higher level of government should have made the best choices. But sub-optimization on discount rates raises peculiar problems. The federal government supports roads, urban renewal, pollution abatement, water supply, etc., with projects designed on the basis of its discount factor. It would not be wise necessarily to substitute these federal projects for local ones simply because of an arbitrarily different discount factor, or to design complementary local projects on the basis of these arbitrary differences.

In the confines of this survey we cannot go further in investigating the issues of appropriate discount rates for urban governments. But we can say with confidence that the planners have oversimplified the problem in their uncritical adoption of the local borrowing rate, and research paralleling the work done on the federal rate would be highly useful. Pending more intensive investigation (and even after completion consensus is unlikely), it would be wise to test out the alternative programs against different discount rates. Some decisions will not be sensitive to the rate. But for those which are, the exercise will be helpful in directing closer attention to the issues.

---

[26]Hufschmidt, et al., op. cit. (footnote 10), chap. II.

Prices

Two general problems are introduced by the necessity of considering future prices in order to evaluate costs and benefits, beyond the omnipresent problems of forecasting. These deal with changes in relative prices in the future which will affect the stream of costs and benefits, and with changes in the relative prices which will come about because of the change in quantity of the public services under consideration. These might be considered two views of the same phenomenon but the problems they pose for analysis are different. First let us say a few words about shifts in general price levels.

In engineering economic analysis associated with public works projects it is common to forecast costs corrected by an extrapolated trend of rising construction and maintenance costs. Though this may result in a more accurate statement of the flow of cash requirements it confuses the analysis for best design. If prospective movements in the general price level that need not necessarily reflect movements in relative real costs were taken into account, the decision makers would be encouraged to substitute current capital costs for future maintenance outlays, or vice versa.

There are, however, shifts in prices of factors which are relevant and too frequently ignored. Relative prices of inputs may change, and in urban works at least that of one input--land--will most likely increase or decrease with relative changes in population, location fashions, etc. Acquisition costs, based on market values, will anticipate such changes where they can be forecast. But what the market cannot anticipate are the large scale changes which a public authority might introduce following acquisition: for example, the addition of 20 per cent to the existing stock of office space through an urban renewal project. For such cases, acquisition costs should not necessarily be a guide to decisions; if they are, there may be excessive or insufficient land reserved for future use. This would not matter so much if land reservations could always be readily changed. But they cannot. Too narrow a road is very expensive to widen and too generous a reservoir is difficult to reclaim.

All this should not discourage an authority from buying land as cheaply as it can in advance of, and generously suited for, its requirements. But the cheap land price itself should not influence the allocation of land to particular uses in the future; the criterion is opportunity costs current at the time of development. "Foresight" should not result in excessive land for reservoirs, streets and so forth.

Shifts in relative prices are just as likely to occur in the case of the products. It is fairly certain that recreational values will increase relative to the general price level and it is necessary to reflect this in the evaluation of benefits. If this were not done we might find the opposite of the consequences predicted in the above paragraph: not enough land devoted to recreation.

The second source of a shift in relative prices is the change consequent on the increase in supply of the service being analyzed. Generally speaking public services are monopolistic and their expansion would involve a reduction in price if they were sold on an open market. In this case the increment in value of product would be underestimated if valued at future prices, but overestimated if valued at current prices. Short of measuring the total area under the demand curve some average of the two prices would be necessary.[27]

"With and Without"

The "with and without" approach may be "instinctive" to economists but it is essentially foreign to government analysis. The latter usually deals with a comparison of the situation "before and after." The important point here is that many things are going to occur independent of the proposal and these benefits or costs should not be attributed to the proposal. For instance, care must be taken that increases in traffic resulting simply from population or income increases are not assigned to an improved highway layout.

But the forecast of trends is sometimes laborious to make and often very speculative. In the instance just quoted, there must be a large margin of error in forecasts of future urban traffic arising from prospective changes in income in the different income brackets. Where such forecasts can be avoided without damage to the analysis, they clearly should be. They might, for example, be ignored when two competitive alternatives will be subject to identical trends on the "without" basis, and it is only necessary to establish the differences between their costs and benefits.

Interdependencies and Externalities

The interdependencies of supply among public agencies and the indirect effects of public services on third parties are among the most difficult problems. The road program directly affects the fire program which affects the water program, and so on. Independent decision making by each agency is bound to create inefficiency. Similar repercussions occur in the private sector and though often they consist of "mere" monetary transfers among units, at other times they involve substantial real gains and costs.

It must be accepted that an already complex analysis is made much more complex if projects cannot be isolated; and that ideally all interdependencies should be considered--except where any program of projects (and therefore by definition any particular project) is small enough not to make much influence on the rest of the economy or the particular sector being considered. Three possibilities can be suggested.

1. Rules reflecting higher-level "external" effects can be given to "lower-level" decision makers. For instance a city's waste treatment can be affected by state or federal laws, penalties or subsidies which are imposed with an eye to its polluting effects. The city analysts need not consider the effects of its treatment alternatives on other communities, since these effects are communicated to them directly in the form of prices or orders. This does not avoid the necessity to study external effects, but it does remove it from the need of lower-level organs to study complex interactions beyond their capacity.

---

[27] McKean, op. cit. (footnote 1), chap. 10.

2. Where nonetheless interdependencies at the lower level must be considered, the aim should be a model which can be handled by mechanical aids to computation. This stage has been reached in various transportation studies where each road project must be considered as part of the total system.[28]

3. Where interdependencies are too complex for this, then the principle of aggregation might be used.[29] In this the program sector might be analyzed in detail while the "rest of the economy" is treated more simply at a macro-economic level. This, in effect, is what is done by accepting prices as datum, for prices are the consequences of the operations of all of the market of the economy.

The external or spillover effects are less tractable. On a general level they have been very well described in the water resources literature. Undoubtedly we shall have to go through the same searching analysis for each of the public services. Without this exploration of the indirect processes, double counting and neglect of significant benefits and costs will prevail. For instance, in the case of a street program the major benefits and the reduced travel costs accrue to the street users. But some analysts might add to that the increases in property values. This would be at least partial double-counting, since the increase in property values is just a capture by the property owners of some at least of the benefits of reduced travel costs. Or to illustrate a possibly neglected feature, we might have increased disease associated directly with the fumes and noise of a larger volume of traffic.

The major difficulties in avoiding muddle in this area is the resistance to allowing offsetting gains and losses to cancel each other out; the focus on dramatic losses or gains in values of existing assets and then the neglect of the less obscure gains or losses to the general public; the understandable confusion between marginal and total. The last problem is to the heart of the difficulty. It is reasonable to argue that without our educational public services our complex industrial structure would be unlikely. (If education were not public, there would be private educational developments which would supply these industries, but the source of the supply is irrelevant here.) Therefore it seems reasonable in measuring educational benefits to add to the increased income of the workers trained at school the profits of entrepreneurs, and so forth. But the question posed to the decision maker is not whether to provide education, but instead whether to add another year to compulsory schooling, or whether to slightly alter the pupil-teacher ratio, or whether to introduce fringe services like counseling. For many of these decisions the outcomes are only marginal to the industrial structure of the economy. Effects on profits and the industrial structure are small and may be safely neglected.

Though it is likely true that the bulk of municipal decisions may be concerned with marginal effects, difficulties arise because in some cases the margins are quite large. A bridge, an expressway, an extension of the

---

[28] E.g., Chicago Area Transportation Study, Penn-Jersey Transportation Study.

[29] Jan Tinbergen, The Design of Development (Baltimore: The Johns Hopkins Press, 1958), p. 81.

school-leaving age by one year, a decision to replenish ground water storage with reclaimed sewage, the motorizing of the police, etc., proceed with marked discontinuities. Secondary effects will not be small, benefits and costs may be found far distant from the direct users, and the effect on supply prices cannot be ignored. Therefore we shall have to develop the appropriate framework for the analysis of spillover effects for many public services, though for the bulk of decisions they will prove unnecessary.

Parties: Balance Sheet or Index

The benefit-cost analyses of resources development projects have been concerned with the computation of a single number to represent the net increase in national product due to the public investment. At least two major issues of evaluation remained outside of this index: intangibles (which are described by the text) and the incidence effects (who is going to incur the cost and receive the benefits). Though these two aspects of evaluation are important in even the commercial activities of multiple-purpose water projects their significance there is small compared to what it is in urban public services, where sometimes they may be the dominant factor. Many may deplore the dominant role of the architect in the city planning profession, but his presence attests to the importance of the "aesthetic" in the public's view of how they want their city organized. Many economists may urge that distributional effects are best handled by lump sum transfers, but public morality rejects this avenue and instead accepts the indirect tinkering of the market place to affect these transfers. The impersonal market place is not only a place where conflicts can be resolved without personal bitterness and violence, but it is also a place where aid can be received without shame or humbleness.

At the beginning of this section we suggested that monetary values can be imputed to the intangibles, and higher weights can be assigned to incomes received by special groups. These are necessary to make the benefit-cost figures sufficient for a decision. But it would be naive to believe that we are close to this stage for most public services. At this stage of science, in addition to those consequences which we can value with a common pecuniary measure, we must delineate the consequences which can be measured only in physical terms, and for some variables only with statements of direction of change, and for some with a qualitative but precise description. Further, since the policy makers are unlikely to assign explicit weights to incomes of different groups they will go outside the measured indices to find bases for decision. Therefore we suggest that the form of the presentation of the analysis should take a more humble approach than the simple presentation of a summary index. It should specify the benefits and costs of all parties.[30]

A tabular presentation of the benefits and costs accruing to all of the parties affected by the decision will not only be an orderly way of assembling all of the relevant data, but it also confronts the decision makers with the hard choices of implicit or explicit assignment of weights to intangibles or favored groups.

This brings us to the measurement of benefits. Since almost every decision involves a balancing of advantages and disadvantages, where

[30] Nathaniel Lichfield, Cost Benefit Analysis in Urban Redevelopment, University of California Real Estate Research Program Report No. 20 (1962), chap. 2.

usually the consequences of any decision are a set of advantages and disadvantages which are realized to varying degrees, measurement is a central feature of successful benefit-cost analysis. Therefore the widespread adoption of these techniques will be as dependent upon the devising of methods of measurement as they will be on the clarification of concepts and institutional reforms. But useful analysis need not wait until that happy day when all of the elements for successful application are present. The structure of the argument of the analysis provides a basis for ordering whatever information is available. The selection of the relevant concepts and their ordering in a meaningful structure is itself a worthwhile venture for urban government decision makers. This might appear dubious to economists, for an essential element of economic analysis is its focus on quantitative factors, and when these are difficult to estimate the analysis tends to be discarded. But even without full measurement, the analysis will still be superior to an argument which contains a recital of irrelevant facts and hunches. Take the analogy of the firm. The most successful use of economic analysis is in those areas where measurement is most feasible and concepts clear, e.g., inventory control and production planning. But it is also proving useful in the decisions dealing with more obscure problems such as investment, location, and organization.

TYPICAL ANALYSES SUGGESTED BY BENEFIT-COST CONSIDERATIONS

Urban governments analyze their decisions by various methods. Such analyses could, we believe, be improved in many cases by borrowing from benefit-cost analysis. By way of example, we shall cite four typical kinds of urban government analysis, based on readily available studies that illuminate particular aspects of benefit-cost analysis. The following aspects are dealt with: (a) effect of a constraint upon choice; (b) criterion for choice; (c) balance sheet summarizing total costs and benefits; and (d) weighting of conflicting goals in making a choice.

Constraint upon Choice: Cost-Revenue and Density

Any new building development will involve a local government in municipal operating costs according to the services provided, and revenues from real property taxes according to the assessment values. Therefore the municipality must take an attitude to the density of proposed residential development, since variations could affect both its cost and revenues. Useful data on such variations exist in an unpublished study relating to Philadelphia,[31] which compares costs and revenues for developing new residential areas at three possible densities, as follows:

[31]William H. Brown, "Redevelopment Decisions: ultimate land uses and their influence on operating costs" (unpublished, undated). Many other good studies exist, but are not as helpful as this in relation to density. This study, it should be noted, adds to the criticism of typical cost-revenue analysis.

	Detached houses	Row houses	
	A	C	D
Persons per dwelling	3.08	3.4	3.2
Persons per acre	10.9	38.7	60.5
Dwellings per acre	3.5	11.38	19.0

The costs in the study are the incremental operating costs for ten city services which would be directly associated with the development, so excluding those which might arise outside the residential areas, such as increased traffic on nearby roads. The ten services are streets, fire protection, fire hydrants, police, recreation, library, public health, welfare, state welfare, and schools.[32] The study does not discuss revenues. For each density, however, it gives the median value of the dwelling units, and this for our purpose is taken as proportional to the respective assessments of the three areas for real property tax.

The following table summarizes the estimated comparative costs and revenues per acre and dwelling, with density A as 100 in each case.[33]

Table 1. Residential development at three densities--indexes of municipal costs and revenues

Scheme and dwelling per acre	Average costs		Medial revenues	
	per acre	per dwelling	per acre	per dwelling
A - 3.5 d/a.	100	100	100	100
C - 11.38 d/a.	350	108	208	64
D - 19.0 d/a.	590	108	262	48

The table shows that in this instance marginal costs per dwelling remain roughly constant, whereas marginal revenues per dwelling decline significantly as the density increases. It follows that on the typical cost-revenue criterion, the ratio of total revenue over cost, the municipality would tend to opt for the lower densities. But the lower the density and greater the dispersal, the higher are costs which are not included in the municipal revenues: state highway system, transportation costs, land costs. The optimum position for the community is thus not necessarily achieved on this cost-revenue criterion.

The table also brings out the effect of one common constraint on choice: land shortage for settlement through perhaps reasons of topography or municipal boundaries. Given this constraint and the objective of accommodating a given number of people, the choice must clearly be for the higher density, and an adverse cost-revenue situation. Put another way, as the Table also shows, with a fixed number of acres, as density increases, marginal costs per acre rise much more steeply than marginal revenues per acre.

[32]In fact, fire protection is shown at zero because it was found that none of the new developments if introduced, at the size considered, has any measurable effect on the fire department's operating costs. A figure should be included for our analysis, but it would not make a significant difference.

[33]The comparative costs per dwelling and costs per person would not be constant because the persons per house varies. But this is an elusive and fluctuating figure, and therefore cost per dwelling is taken instead.

Criterion for Choice: Fire Prevention

In New York City in 1950 there were 44,370 outbreaks of fire resulting in 124 deaths and loss of property estimated at $19.5 million. To meet the situation the Fire Department has an operating budget of $53 million, exclusive of pensions.[34] It was to achieve reduction in costs while maintaining current efficiency that the 1953 Report on Management made an exhaustive analysis of the service.

The actual operating costs of the Fire Department were taken as the cost of fire prevention. But to this might be added costs of combating carelessness. The New York report brings out the large proportion of the damage due to smoking and carelessness with candles, and children playing with matches; no less than 42 per cent of the lives and 29 per cent of the damage. In addition might be added the costs of construction occasioned solely for the purpose of fireproofing, the cost of fire fighting arrangements made by private bodies, the additional cost of car parking made necessary by the prohibition of parking at hydrants.

The benefits from the service are the potential reduction in damage to life and property. The issue that arises is whether additional benefit can be earned by extra cost on the service, or by improvement in efficiency at the same or lower cost. The importance of improving efficiency is seen from the facts. In New York, a mere 0.1 per cent of the fires led to 30 per cent of the damage, 5 per cent led to 88 per cent and 20 per cent to 99 per cent; and most of the 124 lives lost were from smoke which grows as the fire grows. The objective of the service is thus the stamping out of all fires as quickly as possible, for each one carries the potential menace of city destruction. This requires a service capable of dealing with several outbreaks at once; that is, peak conditions.

The approach in the 1953 Report was to review means of saving cost while maintaining the same level of efficiency. The cost savings were not fully itemized but would amount to several million dollars. The level of efficiency was determined by adhering to accepted standards and by testing the proposed new system by reference to the three most critical incidents which had occurred over the past ten years. Perhaps it was this cost minimization approach that precluded an optimizing solution. The question that was not posed was the extra cost necessary for reducing still further the life loss and damage. It was envisaged that a general location of fire companies to conform to an ideal distribution, one presumably which could have achieved this objective, would be beyond the city's resources. This presumably means that it would not have produced an adequate return from the annual property and life loss. This, however, was not established. But even if no major capital investment or relocation was feasible, presumably extra manpower and equipment at given locations would have produced some improvement; and since the annual loss was large when compared with the operating budget there was scope for improvement of results in this direction.

It could not be expected that a review of New York Fire Department would raise the question of the abandonment of the service altogether, but this is certainly relevant in smaller towns and poorer communities. In

---

[34] Modern Management in the City of New York, op. cit. (footnote 8), Vol. II, pp. 709 ff.

fact, many operate without a professional service, implying that its cost is considered to be greater than the potential benefit: the saving from chance destruction of the whole settlement. The damage that would then result is taken to be smaller in price than the cost of running a service which could reduce this towards zero.

Balance Sheet of Benefits and Costs: Preservation of a Historic Building

A complex decision often facing public bodies revolves around the case for preserving particular buildings of architectural and historic interest which are threatened with demolition under the market process. The values maintained by preservation are intangible but the costs are measurable: how can the two be compared? Lichfield has applied benefit-cost analysis to an instance of this kind.[35] It is related to a decision facing the federal government in San Francisco, as owners of such a building, but is relevant to urban government in any city.

The building in question is the Old Mint, covering a one-acre city block, a valuable site, in the central business district of San Francisco. Built in 1869 it has not been used for minting since 1937 and is suitable only for somewhat inferior offices for government staff (National Park Service) if reconditioned at considerable cost. This was one alternative facing the government. The other was demolition, and accommodating the National Park Service in a new Federal Building to be erected in the San Francisco Civic Center.

In tracing the repercussions, the analysis considered not only the implications for the federal government but of eight other sectors of the community who would be affected. This was summarized in two tables which are reproduced here, illustrating the balance sheet mentioned above. The parties are grouped into "producers" and "consumers" of services on Table 2, the producers having an odd number on the left hand side and the appropriate consumers an even number on the right. For each party it traced through the total costs and benefits under each alternative. Since so many intangible, and measurable but unmeasured, entries are included in the Table its implications are difficult to grasp. Table 2 is therefore summarized in Table 3. This gives the net difference in costs and benefits if the Old Mint is retained rather than demolished, in a form which enables all implications to be readily comprehended.

The tables and supporting data show that without rehabilitation, the Old Mint could be sold for $700,000; and that the investment of $1,000,000 in rehabilitation would add only $630,000 to its value, producing a total loss of $370,000. They also show that to rehabilitate and use the building would be a poorer investment than providing for the National Park Service in the new Federal Building. For the latter the site would cost nothing since no extra land would be needed; and the investment of $1,350,000 in building would produce a market value of $2,500,000, to which would be added the proceeds of selling the Old Mint site, $700,000, a gain of $1,150,000. Thus if there were no constraint on capital expenditure, it would pay the federal government to sell the property and accommodate the Park Service in the new building.

---

[35] Lichfield, op. cit. (footnote 30), chap. 3.

Table 2. Benefits and costs for alternative uses of Old Mint

$000

Item No	PRODUCERS	1 Conditions Existing				2 OM for NPS Offices				3 OM demolished & sold				Remarks
		Benefit		Cost		Benefit		Cost		Benefit		Cost		
		C	A	C	A	C	A	C	A	C	A	C	A	
1	Federal Govt as Property Owners													
	a) Existing NPS Offices		m_1		m_1									
	b) Old Mint – Initial – Property									700				
	Restoration							1,000						
	Recurring – Income						180							
	Operating				30				85					
	c) New Fed Building, NPS Section													
	Initial											1,350		
	Recurring – Income										225			
	Operating												100	
	Reduction				30		65				106			
3	Nearby Property Owners													
	a) Visitors		m_2				m_2+							
	b) O.M Site Employees						m_3				m_3+			
	Reduction		m_2				m_2+ m_3				m_3+			
5	Other Property Owners													
	a) Existing NPS Offices		46		19				12				12	
	b) Floating Development Value	1,000				1000				E				1
	Reduction		67				28						12	
7	City Operation – Tax revenue & costs													
	a) Existing NPS Offices				m_4				m_4				m_4	
	b) Old Mint				m_5				m_5		150		m_5	
	c) New Federal Building				m_6				m_6				m_6	
	d) Floating Development Value		m_7		m_8		m_7		m_8					
	Reduction		m_7		m_8		m_7		m_8		150			
	Adjustment						$\underline{m_9}$				$\underline{m_9+}$			
	Final Reduction													
9	City Economy													
	Tourists and Students		m_{10}				m_{10}+							
11	Fixed Capital Investment													
	1b) Old Mint restoration							1,000						
	demolition											300		
	new building											M		
	1c) New Federal Building											1,350		
	5b) Floating Value							M–						
	Total							1,000 M–				1,650 M		

Item No	CONSUMERS	1 Conditions Existing				2 OM for NPS Offices				3 OM demolished & sold				Remarks
		Benefit		Cost		Benefit		Cost		Benefit		Cost		
		C	A	C	A	C	A	C	A	C	A	C	A	
2	National Park Service													
	a) Existing Offices		i_1		46									
	b) Old Mint						i_2		180					
	c) New Federal Building										i_3		225	
	Reduction													
4	Nearby Business Proprietors													
	a) Tourists		m_2		m_2		m_2+		m_2+					
	b) O M Site Employees						m_3		m_3		m_3+		m_3+	
	Reduction													
6	Other Occupiers													
	Reduction													
8	City Taxpayers													
	Reduction								$\underline{m_9}$				$\underline{m_9+}$	
10	Public													
	a) Citizens, Tourists & Students		i_4		m_{11}		i_4+		m_{11}+					
	b) National Heritage		i_6				i_6							
	Reduction		i_4 i_5		m_{11}		i_4+ i_6		m_{11}+					

Notes to Table 2

1. All columns show absolutes.
2. All figures are in $000.
3. "C" is a capital (once for all) item; "A" is an annual (continuing) item.
4. "M" or "m" shows a measurable item which has not been measured, initial or annual respectively. In each item different numbers (m_1, m_2, etc.) indicate different figures.
5. "I" or "i" shows an intangible (nonmeasurable) item, initial or annual respectively. The numbering (i_1, i_2) is as for "M" items.
6. The entries which are underlined are negative.
7. The reduction shows the annual net benefits and costs under each item other than capital investment.

Table 3. Old Mint. Summation of Table 2--difference in annual costs and benefits that would arise if Old Mint were retained and occupied by N. P. S., instead of being demolished.

$ 000

Item No.	PRODUCERS	Scheme 2 minus Scheme 3			Item No.	CONSUMERS	Scheme 2 minus Scheme 3		
		Ben.	Cost	Net			Ben.	Cost	Net
1	Federal Govt. as Property Owners	$\underline{41}$	·	$\underline{41}$	2	National Park Service	·	·	·
3	Nearby Property Owners	m_2+ m_3 $\underline{m_3+}$	·	m_2+ m_3 $\underline{m_3+}$	4	Nearby Business Proprietors	·	·	·
5	Other Property Owners	28	$\underline{12}$	40	6	Other Property Occupiers	·	·	·
7	City Operation	·	·	·	8	City Taxpayers	m_9 $\underline{m_9+}$	·	m_9 $\underline{m_9+}$
9	City Economy	$m_{10}+$	·	$m_{10}+$	10	Public	i_4+ i_6	$m_{11}+$	i_4+ i_6 $\underline{m_{11}+}$
	Total net			m_2+ m_3 $\underline{m_3+}$ $m_{10}+$		Total net			m_9 $\underline{m_9+}$ i_4+ i_6 $\underline{m_{11}+}$

Notes to Table 3

1. The benefits and costs shown are obtained by deducting the Table 2 reduced items in Scheme 3 from Scheme 2, keeping benefits and costs separate in the columns so headed. The net is then shown in the third column by deducting costs from benefits. A + benefit and - cost are +; and a - benefit and + cost are -.
2. Where M and I items do not permit of arithmetical reduction in the net column, probabilities are forecast.
3. All quantities are in annual terms.
4. Figures are in $000.

As to the responsibility of the government as custodian of historic monuments, no computation could be made of the value of the Old Mint to the nation. But the federal government could calculate the cost to itself of retaining the value. This would be the loss on foregoing the more profitable of the two courses of action just outlined. The loss would be $2,220,000, for the rehabilitation of the Old Mint would result in a loss of $370,000, and the accommodation of the Park Service in the new Federal Building would result in a gain of $1,850,000. This is the price the nation would have to pay (in taxes) to buy for itself the Old Mint in perpetuity. The two issues would be sufficiently defined for the federal government to make a decision, considering only its direct costs and benefits.

But this does not include the total costs and benefits which are shown in the tables. First, the government would know that while it would lose by the amount stated to preserve the Old Mint, other property owners would gain. Second, when under pressure from the city to demolish the building in order to release a site which would contribute to the tax roll, the government could point out that the actual loss to the tax roll is considerably less than the tax potential on the Old Mint site if redeveloped. The building which would have been erected on the Old Mint site would, in response to demand, have been built elsewhere to earn taxes for the city. Furthermore, it could point out the retention of the Old Mint would help the city economy to some degree. These considerations should remove any doubt that the Old Mint would be worth preserving, if the federal government wished to base its decision on total costs and benefits to the community and not merely on its own.

Balancing Conflicting Goals: Alternative Plans for a Central Area

We suggested above that the formulation of goals by public bodies might take the form of identifying the dimensions of utility or welfare, and that choice involves selection among various packages of such identities. This process has been explored by Margolis, in a preliminary model for a cost-benefit analysis of a very complex decision facing a city government: on the alternative city plans for the central area of a British university town.[36] And it is fitting that we close with what we have recognized as the most perplexing part of the analysis: the goals.

The issue is this. In this city of about 100,000 people the university and central business uses are complementary, but are expanding, in competition for space. The satisfactory resolution of the competition has been a live issue for some fifteen years. The initial decision in a plan of 1950 was to restrict the population size of the city so that the central business district need not grow in detriment to the university character of the center. But despite such population restriction, the center shows signs of unexpected growth because of rise in regional population, purchasing power and car usage. To avoid destruction of the university character, a change in plan is advocated: a new central business district should be provided by residential redevelopment about one mile away and the current center be allowed to lapse into a service area for the University and tourists. The question arose: what are the comparative costs and benefits of the alternative plans?

[36] Julius Margolis, "Two Alternative Plans for a Central Area: a preliminary model for a cost-benefit analysis" (unpublished).

An essential preliminary of the analysis is to bring clarity into the definition of goals as a basis for choice. What are the qualities which must be sought in a University town?[37] What sacrifices must be made to retain them? Are they incompatible with other conventional goals of urban living? First, the sectors of the community which would be affected by the proposals were enumerated. These were eight in number: landowners, developers, University, traders, shoppers, through motorists, local taxpayers, the nation. Then for each were listed their private goals and the instrumental objectives in reaching these goals. The next step was a social evaluation of these goals: which should the city adopt as in the public interest? The analysis proper then began with a statement of the nature of the costs and benefits which would fall on each of the sectors under the alternatives and an indication of the method of measurement. But not all these costs and benefits are relevant to the public interest, so that a selection based on a social evaluation of their incidence was then attempted. From this the next step would be the measurement of the socially significant costs and benefits which would arise, or the qualitative description where measurement is not practicable. Judgment would then be made on the basis of the prior social evaluation of goals.

In such judgment there would be a need to balance the implications of the alternatives in respect of conflicting goals. The following extract from the Margolis Report indicates an approach to this:

> Three goals have been identified: an efficient trading centre which serves a residential population of 100,000 and a regional hinterland of an equal number; the enhancement of a University with its complex educational processes and needs for laboratories, dormitories, lecture halls, recreation grounds, and so on; and the conservation of a historic site which produced a happy conjunction of energetic trade in confined quarters and monumental architecture of the Colleges. None of these goals can be fully pursued without conflicting with the others since all of these activities infringe so closely on the town centre. Instead the task of the Plan is to select a target in terms of trade, accessibility and University functioning which somehow is optimal, i.e., the value of the loss to the University's functioning because of the Plan's conservation of the historic image and provision for efficiency should be equal to the loss of efficiency because of the Plan's preservation of the historic image and University efficiency and in turn equal to the value of the loss in the preservation of the historic image because of the Plan's provision for efficiency and University functioning. There are no "scientific" rules by which the balance of goals can authoritatively be derived. Judgment, wisdom, goodwill are critical in the evaluation. But there is a role for systematic analysis, for calculation of "imponderables," and for the testing of feasibility and consistency which would be of considerable help in the decision process.

---

[37] The criteria for deciding just what is a true university town has been explored by Edmund W. Gilbert, The University Town in England and West Germany, Research Paper No. 71, Department of Geography, University of Chicago (Oxford: Dittons and Blackwell, 1961).

The implications of the plan for the three goals must be handled quite distinctly. There are techniques by which one could try to compare a unit of University functioning with a unit of trading efficiency but such explicit value weighting is not on the agenda today. For the University we require a statement of their Plans for their own operations, why the elements of the plan are necessary, what alternatives are possible; and wherein do the rejected alternatives fall short of the desired choices. Therefore for the University "needs" we do not have a quantitative scale of satisfaction, i.e., a measure of efficient alternatives. But we can quantify many of the factors which do enter into their determination of an optimal plan for their operations and the degree of inefficiency in rejected alternations, e.g., frequency of contacts of faculty of different departments; time spent in traveling to lectures, to library, to meals, to recreation, . . .

The preservation of the conjunction of town and gown is a much more difficult goal to identify and quantify. There is no institutional group which can specify its satisfaction or plead for its character. One way to approach the problem is to attempt to identify a desirable ratio of University to non-University personnel in the centre; or a ratio of University to non-University shoppers on the streets, in the restaurants, . . . Or possibly one can identify a minimum mass of non-University contacts necessary to make an impression on the Centre, and the Plan should seek to achieve a non-University traffic between this minimum level and a level which would raise the threat of possible engulfment of the University.

The goal of an efficient trading city is more amenable to quantitative analysis but this boon is offset by the difficulties of forecasting the technology and preferences of the distant future. We know the historical site that we want to preserve though we may not all be agreed about its virtues. Educational and research needs will change, but not extensively, But the technology of retailing, the means of transportation and communication, the preferences of shopping households are all going to change dramatically in the coming half-century. Some of these dynamics are already apparent and have begun to put a heavy strain on the structure of the centre, e.g., the private auto. But other changes, which today may be only fancies, will be initiated in the future which is to be constrained by our Plan and the Plan must have sufficient flexibility to account for this unknown.

The efficiency of a trading city is judged in the market place but when a Plan does not allow the full working of the market then evaluation becomes indirect. An operating criterion for an efficient organization is that the alternative should be chosen which gives the participants the greatest difference between the price they are prepared to pay for the organization and the costs of making it available. Without the free functioning of a market it becomes difficult, though not impossible, to judge the "willing price" and even a proper estimate of cost. Since this one goal is conceptually quantitative it enables us to ask the question "How much of a loss in efficient working are we prepared to absorb in order to enhance University efficiency or in order to achieve a 1 per cent increase in non-University central traffic above the minimum?" The answer to this question enables us to assign a "cost" to these other goals which makes them all commensurable. But this task is probably beyond the limits of this modest investigation. We shall have to be satisfied with an analysis of each of the three dominant objectives as independent dimensions and leave their comparisons to more intuitive bases.

The three goals have been formulated in a spatial structure. In the first round of analysis the efficient University and the historic image are fixed targets given to us by University planners and social consensus while the goal of an efficient trading city is open-ended. That is we seek to maximize the efficient trading city subject to the constraint that certain targets in regard to University functioning and the preservation of a historical balance are achieved. In the second round of analysis we may find it necessary to amend the University and historic goals if we find that they involve an excessive loss in efficiency. And of course in the first round of specifying the University and historic targets these have been formulated so that they are consistent which probably means some sacrifices of each goal.

7

Costs and Benefits from Different Viewpoints

by Roland N. McKean

When people live in groups or even have any significant contacts with each other, it is clear that they have to do some planning--that is, they have to think about the implications of alternative rules and arrangements and choose among them. People may devise rules that permit considerable decentralization and individual choice, or they may have rules under which most issues are settled by authority or tradition, leaving less scope for individual choice. In any case, these rules constitute planning: they are not chosen at random but are the result of some sort of political process in which various persons did some thinking (and also a good deal of compromising). In an urban area, the existing complex of laws, ordinances, and institutions for governing constitute a kind of "urban plan"--usually a rather helter-skelter one but nonetheless the result of human planning as we know it.

All too often this process--indeed the exercise of central authority in general--has led to planning of the people, by the few, and for the few. Also, urban plans in the formal sense have too often been chosen in terms of narrow partial criteria--efficiency in carrying out one function regardless of undesirable effects elsewhere, esthetics alone without regard for other consequences, or the betterment of one group regardless of the implications for others.

It is reasonable to believe that we can do better, and in recent years, there has been growing interest in the use of broader criteria for the analysis of city plans. There has been growing recognition that criteria used in the past have included only part of the things desired and part of the sacrifices entailed. Hope has sprung up of using cost-benefit analyses that take into account more of the costs and more of the benefits of alternative institutional arrangements.[1] Of course, we shall never be able to include, at least quantitatively, all of the relevant costs and gains. For example, we cannot measure adequately the significance of a plan's impact on the range of individual choice or on the probability of maintaining individual rights. But we can hope to measure more of the costs and gains than were considered in the past and measure them more accurately than in the past, making final evaluations easier (though still not easy by any means).

In devising and evaluating rules for social organizations, the forces that raise the major problems are the discrepancies between individual costs (or benefits) and our conceptions of total costs (or benefits). If no discrepancies existed, each individual could do whatever he pleased, and we would have no objection. These discrepancies are the main reason we have police departments, zoning ordinances, or any form of urban planning. They are also the reason for wanting relatively broad analyses of alternative city plans, because fairly comprehensive cost-benefit analyses are necessary to reveal the indirect effects as well as the direct ones, the costs and gains in total as well as to selected groups or individuals. The basic difficulty is that the costs and gains felt by one group are different from those perceived by another group. The cost and gains felt by the mayor are quite different from those felt by citizen A or B and still different from our conceptions of total costs and gains.

But our problems would not be solved even if cost-benefit analyses could reveal the plan that in the abstract would yield the greatest total benefit (net of costs). The problems would not all be solved because these same divergences between individual interest and total costs and gains are likely to thwart the implementation of many plans. In trying to put rules or plans into effect, we must think very carefully about ways to cope with the resulting pressures. Moreover, in many instances we should apply a a "degradation factor" to the net gains promised by a plan to allow for the ways in which the results would go awry because of these discrepancies between individual interests and the aims of the plan. In fact the measures that seem second- or third-best in terms of straightforward cost-benefit analysis may appear to be best when one reflects on how the various proposals would actually turn out.

These statements are simple propositions about human behavior and political realities, but there is danger of losing sight of them as analyses become more sophisticated and more dependent upon the skills of physical scientists, economists, architects, and others who are not always vividly aware of the nature of political processes. Because of such danger, it is worthwhile to examine the reasons that these simple propositions are so important.

SELF-INTEREST AND THE IMPLEMENTATION OF URBAN PLANS

When people appraise proposals for government action, they often have extremely misleading models of political and administrative behavior

---

[1]Nathaniel Lichfield, "Cost-Benefit Analysis in City Planning," Journal of the American Institute of Planners, November, 1960, pp. 273-79. See also Lichfield's Economics of Planned Development (London: Estates Gazette, 1956).

Harvey S. Perloff traces out the way urban planning has evolved in Education for Planning: City, State, and Regional (Baltimore: The Johns Hopkins Press, 1957), especially pp. 9-24.

in mind. Some persons assume that every politician or official is vicious, having as his primary aim the achievement of evil. More persons are at the other extreme, I would guess; they slip into assuming that anyone paid by the public will act in the interests of the public--that any activity placed in the public sphere will automatically be conducted in the public interest. Yet a little reflection ought to remind us that this is not necessarily so. Consider the following comment on an activity well below the policy-making level:

> At first sight there hardly seems to be a problem here. If we employ a man to direct activities at a playground, *of course* he will spend his time each day from nine to five at the playground; *of course* he will organize and direct play activities there; *of course* he will carry out the policies formulated by the directors of the organization who are responsible for planning its program. Only the many instances of organization failure--instances where an organization does not carry out its task or where it succeeds at an excessive economic and human cost--warn us that there is really nothing automatic about the process.[2]

The two extreme models of government officials' behavior--that government action is always evil and that public activities are automatically conducted in the public interest--are dangerously unrealistic. No model of behavior can be 100 per cent realistic, of course, but these extremes are too far from the mark for them to serve us usefully. Actually, government officials and politicians are much like business employees and administrators. They are not a random sample of the population, for they have above-average ability and ambition, but they are not a separate breed. As for other citizens and voters, who also play vital roles in political processes, their motivation is fairly understandable. None of these groups is ceaselessly trying to do evil, on the one hand, or to secure the maximum total benefit for the nation, on the other.

In general terms, an individual is moved to act by anticipated costs and gains *as he feels them*. Sometimes the terms "deprivations" and "gratifications" or "sacrifices" and "satisfactions" are used instead of costs and gains, but the idea is the same. In this general form, this kind of proposition is a tautology. It is about like saying that a man acts because of the things that make him act. But discussions of behavior can start out with this framework and lead to significant propositions. With the aid of any insight into what factors cause individuals to feel gratification or deprivation, one can arrive at a predictive model rather than an empty tautology.[3]

What *are* the principal items that affect the gains and costs felt by an individual? Material comfort for himself and his family is clearly a major item. Some persons believe that economists regard material gain

---

[2]Herbert A. Simon, Donald W. Smithburg, and Victor A. Thompson, *Public Administration* (New York: Alfred A. Knopf, 1950), p. 55 (italics in original).

[3]A number of persons have contributed extremely useful analyses or discussions. For example, see James G. March and Herbert A. Simon, *Organizations* (New York: Wiley & Sons, 1958), especially pp. 9-11 and 83-111; Robert A. Dahl and Charles E. Lindblom, *Politics, Economics, and Welfare* (New York: Harper & Brothers, 1953), pp. 93-117; and Anthony Downs, "An Economic Theory of Political Action in a Democracy," *Journal of Political Economy*, April, 1957, pp. 135-50.

or loss as the only factor motivating man. They believe this because economists have talked a good deal about "economic man." This term does suggest a person with a one-track mind, but in point of fact it has usually meant "rational man"[4] as distinguished from "neurotic man." It has not referred to a person who pursues only material wealth.

Material wealth and comfort, then, is one of the desiderata, but there are many others.[5] Just where to draw the line, nobody knows, but one surely gets more insight into average behavior by confining the list to major and rather obvious driving forces--such as the desires for prestige, material gain, power, security, and avoidance of difficult decisions and inconveniences--and by ignoring the minor or bizarre factors that may be operative. By the same token, we would do well for the most part to neglect completely neurotic behavior in trying to predict performance (though we should certainly keep the threat of neurotic behavior in mind when appraising plans for cities or other social organizations).

We should keep in mind too that probability of success figures importantly in calculating the expected reward from undertaking some action. No matter how much a government official values personal promotion or civil rights, he will not be moved to take an action to further those aims if the action has a near-zero probability of achieving those ends. Each of us has an enormous stake in preventing thermonuclear war, but few of us put in several hours per day in an effort to prevent it. Why? Because we know that individual efforts of ordinary persons can have almost no influence on the outcome. This is another reason that many motives that might be called "noble" ones play so small a role in shaping our actions.

One major question about costs and rewards is the following: For whom does a person want increased gratifications and decreased deprivations? If he wants good things for everyone with equal fervor, then propositions about his motivation become very general again, and can explain any sort of action but predict none. It is fairly clear, however, that most persons want the desirable things primarily for themselves and their immediate families. To a lesser yet significant extent, they are concerned about the wellbeing of close friends. If we look outside this circle, the amount of genuine concern falls off rapidly. (Anonymous philanthropists do exist, but they are the exception rather than the rule.) Once more, we don't know just where to draw the line. But I submit that we get considerable insight into behavior if we adopt a sort of "cookie-cutter technique"--if we assume that each individual is concerned about himself, his immediate family circle, and a few close friends, and neglect whatever concern he may feel for other living persons or for unborn generations. (We get no insight at all if we make the cookie-cutter embrace everyone.)

To be sure, each of us identifies with various groups such as the department or organization where we work. A gain for the Republican party is a gain to the individual members. A blow to the firm is a blow to the employee. Even so, these loyalties stem in large part from the fact that the individual has tied his personal prospects, at least for the moment, to that group or organization. Typically one's loyalties and attitudes shift

---

[4]Note that there are severe limits to man's ability to know and to choose wisely, however. Rationality has to mean, not omniscience and optimal choice, but something like trying to do the best one can in the circumstances.

[5]For a detailed discussion of the influences on behavior, see Simon, Smithburg, and Thompson, op. cit. (footnote 2), pp. 55-91.

quickly when he accepts a job with a different firm or moves from one government department to another.

The assumption here is, then, that officials, employees, and voters will fairly consistently act in their own self-interest rather narrowly conceived. That is, each will try to increase gains and reduce costs--in terms of wealth, prestige, power, security, and convenience of working and living--for himself, his family, and to a lesser extent his close friends.[6] This assumption is by no means photographically realistic. People *are* concerned to some extent with unknown persons and abstract principles. Some devote their lives to such causes, and in crises numerous individuals have risked death to save total strangers from tragedy. Most of the time, though, the rather narrow self-interest assumption does not depart wildly from reality, and I believe it is the kind of abstraction from reality that is useful, the kind of model of human behavior that we would do well to keep in mind.

DISCREPANCIES BETWEEN INDIVIDUAL AND TOTAL COSTS AND GAINS

We must recognize too that ordinarily there are serious discrepancies between the self-interest of the individual voter, employee, or official and the interest of the whole group. That is, there are important divergences between the costs and gains felt by each individual alone and the total effects that cost-benefit analysis seeks to measure. Discrepancies between individual and total costs or gains in the private sector of the economy have been discussed for many years. These are sometimes entitled "external economies or diseconomies" or, perhaps more graphically, "spillover effects." Wherever these effects are important, we try to rig the costs and rewards to individuals so as to induce them to take these spillovers into account. But some persons damn the whole private-enterprise system on the grounds that these external effects are too pervasive.

We must not forget, however, that analogous phenomena are present in the public sector of the economy. It is probably even harder there to bring self-interest into line with community interest. One reason is that many governmental functions are placed in that sector precisely because external effects of those activities are particularly serious, making it difficult for the private-enterprise system to handle them properly. Another reason is that it is in any event difficult in government to utilize markets and voluntary exchange, or any other devices to *reveal* costs and rewards explicitly. It is also difficult to make use of competition and bargaining in such a way as to make the right costs and rewards *felt*, even if they are revealed. Whatever the reasons, though, the governing process is shot through and through with discrepancies between self-interests, other group interests, and total community interest.[7]

Good examples are provided in connection with government proposals to deal with water problems in the Washington, D.C., area. Apparently

---

[6]For public officials, as will be seen, this involves trying to survive or advance in the political-administrative hierarchy, which imposes many constraints on the official's behavior. The pursuit of self-interest does not imply that constraints or costs will be ignored.

[7]See Herbert A. Simon, *Administrative Behavior* (2nd ed.; New York: Macmillan Co., 1961), pp. 63-64, 186-88, 198-217.

Falls Church, Virginia, has tapped some of the best sources of water in the vicinity and is reselling the water at a tidy profit to various cities in Fairfax County. Now the county officials are greatly upset because they must get water from costlier sources than those that would be available if Falls Church had not already tapped them.[8] Another example concerns what is literally a spillover from an anti-pollution proposal:

> Today I am speaking on behalf of the Accokeek Citizens Association, the Accokeek Democratic Club, the Accokeek PTA, the Moyaone organization, the Piscataway Co....
>
> We have joined together in the Potomac Tidewater Council.... This council has one purpose; that is, to reaffirm to this committee that the river does not end at 14th Street.
>
> All the comprehensive plans prior to this committee's studies, work their way down from the headwaters to 14th Street....
>
> The Army engineers in their report did a little better than some. They promised, if we would support their upper river work, to float the sewage down to us a little faster. "Get your sewage while it's fresh, fellows," seemed to be their slogan.[9]

What kind of trouble can arise because of the differences between the self-interest of individuals and the community interest? In general terms the trouble is that wrong actions and policies are undertaken. An individual will take steps that look good from his standpoint even if those actions inflict damages on others. (Unless he has to compensate the others, in which case he is made to feel those costs by "buying" those deprivations from others.) Or, an individual will not take steps that do not look good from his viewpoint even if those actions would bestow significant gains on others. (Unless he is allowed to feel those gains by "selling" those benefits to the recipients.)

In urban planning the result can be inaction if few of the community gains are felt by officials or if officials encounter personal sacrifices that are not real costs from the community's standpoint. Suppose a subsidy is currently being paid to owners of tenements (or any other industry), and an urban plan that looks good in terms of comprehensive cost-benefit analysis calls, among other things, for removal of that subsidy. Council members may feel the gains, e.g., pressure from taxpayers and affected groups, very slightly, but feel the sacrifices (loss of the support of some highly articulate and influential groups) very keenly. The result may be no action at all. The thing to be stressed is that the persons involved need not be acting in any malevolent or anomalous fashion. Most of us, if placed in any of these roles, would behave in much the same way. Costs and rewards from one person's standpoint are different from the costs and gains that are perceived by another, and these costs and rewards are like the

---

[8] Washington Metropolitan Area Water Problems, Hearings before the Joint Committee on Washington Metropolitan Problems, Congress of the United States, 85th Congress, 2nd Session (Washington: U.S. Government Printing Office, 1958), pp. 559-63.

[9] Ibid., p. 338.

strings on marionettes--they keep pulling and affecting behavior in rather predictable ways.

Or, if it is not inaction, the result may be a highly distorted set of measures instead of the recommended plan, or unanticipated responses that lead to results quite different from those planned. These are common outcomes that need no elaboration or illustration. Nearly every city's history reflects the many slips between planning and doing. Again there is little use in abusing the marionettes. The thing to do is to work on the strings to see if we can bring about more desirable outcomes. Sometimes, to be sure, there are despicable individuals and actions involved. The power of self-interest often induces officials to step outside the law, and there is illegal or immoral behavior, rather than merely inequitable, inefficient, or frustrated urban planning. Even in the case of corruption, however, we should recognize it as a product mainly of the cost-reward structure, and we should try to re-rig the strings rather than merely hope for better marionettes. Too often we look upon "the shame of our cities" as a product of bad luck or declining morality and try to use the weak strings of exhortation alone to manipulate the participants.

One key actor whose behavior is often gauged badly is the ordinary citizen affected by urban planning. When laws are made or altered, he will adjust wherever possible so as to reap benefits and avoid losses. He will not usually forego available gains or incur unnecessary losses just to further community objectives or improve the lot of other groups. We must not expect producers or voters or officials to act regularly against their own interests narrowly conceived. We must not expect the farmer facing an acreage limitation to refrain voluntarily from farming his land more intensively. We must not expect citizens to move to locations where they do not wish to live. We must not expect the voter to inform himself and get to the polls when the costs of doing so exceed the gains.[10] (Unless he has a special stake in an issue, the man who votes sacrifices time yet gains only an infinitesimal probability of affecting the outcome.)

There are analogous divergences between individual and total interest at national and international levels. In the relationship among nations, in fact, these discrepancies can and probably will lead to disaster. The total net gains from avoiding thermonuclear war are tremendous. Yet each national leader must look at the gains and costs of alternative defense policies from his own nation's standpoint. And since he cannot control the other nation's behavior, his country may suffer more if he does not run a high risk of war than if he does. The resulting cost-reward pattern as each nation sees it is vastly different from the total costs and gains associated with alternative policies. To reduce the chances of catastrophe, we must assess those cost-gain patterns carefully--e.g., the costs to an enemy of striking first versus the expected costs to him of not striking first, or, as another example, the cost to the U.S.S.R. of permitting suitable inspection arrangements. Furthermore, we must then influence those cost-reward patterns, perhaps through unilateral action or perhaps through enforceable multilateral agreements. We cannot hope for much if we rely on exhortation or mere replacement of the marionettes.

Now, emphasis on conflicting political pressures is nothing new (in either urban planning or planning for peace). Nonetheless, I think it needs

---

[10]Downs, <u>op. cit.</u> (footnote 3), pp. 135-50.

to be further emphasized that these pressures stem mainly from the cost-gain patterns that confront each decision-maker, that self-interest rather narrowly conceived is a powerful force, and that to a considerable extent the social organization must try to harness rather than override individual self-interest (or at least recognize rather than ignore its implications).

COSTS AND BENEFITS FROM DIFFERENT STANDPOINTS

To see more vividly how the gains and costs of alternative decisions must appear to various officials, let us review two cases pertaining to urban planning measures.

Public Housing in Chicago

Consider first public housing decisions in Chicago during the late 1940's. The background and influences at work have been described in some detail by Martin Meyerson and Edward C. Banfield.[11] How did decisions about public housing proposals look to, say, an alderman from a middle-class residential section of the city? To survive and perhaps advance in local politics, he had to please a majority of his constituents, not on each and every issue, but on enough issues to keep their support. Also he had to secure the co-operation of many other aldermen, particularly the more influential ones who were chairmen of important committees, for the Council could ruin him if it really wanted to do so. The Council could do this by withholding patronage or campaign funds, by letting public services in the ward deteriorate, by encouraging investigations or other actions that would embarrass the alderman, and so on. The alderman was dependent upon the Council for many things that were provided to him and his constituents.

Since he received only $5,000 per year in salary, he usually had outside interests, such as an insurance or real estate business. Often, therefore, he was especially concerned about keeping the favor of those politicians or constituents who brought business and outside income to him. He was dependent in some degree upon the mayor, other city officials, ward committeemen (where they were not the same persons as the alderman), state politicians, the press, and perhaps particular ethnic or religious groups. Typically there were things the alderman could do for these persons, and in turn he was partially dependent upon them.[12]

The chain of interdependence was usually a long and complicated one--A would be in some degree dependent upon B because B had some leverage over C, who had some leverage over D, who had some influence with group E, which could threaten or reward A. The existence of these relationships--of the costs and gains confronting A--did not have to be reaffirmed each time an issue arose. Indeed they did not have to be stated explicitly at any time. If A was reasonably intelligent, he automatically asked

---

[11]See their book entitled Politics, Planning, and the Public Interest (Glencoe, Ill.: The Free Press, 1955).

[12]Most books on public administration give the reader some feeling for these interrelationships. A systematic discussion of influences on municipal decisions is presented in Wallace S. Sayre and Herbert Kaufman, Governing New York City, Parts 2 and 3 on "Strategies of the Contestants" (New York: Russell Sage Foundation, 1960), pp. 121-708.

himself: "What are the pleasant and unpleasant consequences for me if I take a particular action?" If he was fairly astute, he could answer the question well enough to survive and advance.[13]

In these circumstances how did the gains and costs of public-housing decisions appear to our hypothetical alderman? Consider, for instance, the costs and gains of voting against a public-housing proposal that included one site in his own ward (we shall call this "Proposal X").

Effects on Alderman A of Opposing Proposal X

Costs

Loss of support of some aldermen and (maybe) of the mayor.
(1) Some felt the general pressure for city to get housing largely at expense of fed'l govt.
(2) Several aldermen favored public housing, especially if site were not in their wards.
(3) Deadline for getting federal money was close.

Loss of goodwill of certain state and federal officials who wanted the program expanded--goodwill that was valuable to the Council and might be valuable someday if not at that moment to the alderman.

Loss of support and insurance business from contractors in ward who would help build public-housing project.

Loss of good relationship with Chicago Housing Authority (but this would not cost the alderman much).

Loss of personal satisfaction in seeing the program materialize (here we assume that this alderman favored public housing in principle).

Loss of support he might otherwise gain from occupants of public housing project (but where their support would go was really quite doubtful).

Benefits

Gain of many constituents' favor. Residents opposed having site in that ward, because public housing believed to lower values of middle-class properties.

Gain of certain key aldermen's support (they opposed the proposal for their own reasons).

Retention of insurance business from pleased constituents and politicians.

Protection of value of alderman's own properties in his ward.

Avoidance of possible racial conflicts and troublesome issues in his ward.

Possible gain of support of city transportation officials, who wanted the site for freeway access roads, etc.

---

[13] The subtle ways in which influences are felt and the fact that interest groups often need no overt representation when decisions are being reached are well known. For instance, see David B. Truman, The Governmental Process (New York: Alfred A. Knopf, 1959), p. 449.

These are just a few of the costs and gains as seen by the alderman, but the list includes several of the most significant ones. Note that the most important costs and benefits from the nation's standpoint (or the city's standpoint) are not on the list--for example, the resource costs of using the sites and building the project, the worth of the improved housing to the probable occupants, or the long-run impact on racial integration. In other words, there may be huge discrepancies between the public interest and the official's interest. Note too that personal principles were probably relatively minor considerations--if the alderman had strong wishes to survive in politics and in his business activities. Personal views can play a larger role, of course, if the alderman (or other official) is independently wealthy or can shift to another occupation with little sacrifice.

Suppose we turn now to the costs and gains of opposing Proposal X from the mayor's standpoint. The consequences with which he had to be concerned were somewhat different.

Effects on Mayor of Opposing Proposal X

Costs	Benefits
Loss of support of many voters (1) Voters, businessmen who would profit directly from the project. (2) Voters who felt the advantages of getting project mostly at federal expense.	Gain of support of voters and ward committeemen who opposed project, mainly in wards containing proposed sites.
Loss of goodwill of certain state and federal officials who wanted the program expanded--goodwill that could affect patronage, campaign funds and support, political prospects of the mayor.	Gain of firmer support from key aldermen in whose wards sites had been selected; and these aldermen might be strong enough to keep Council pretty much in line.
Loss of harmony in Council, support for mayor's programs, backing of certain ward committeemen in next election.	Avoidance, or postponement, of extending areas of racial conflicts.
Loss of good relationship with Chicago Housing Authority, through which Chicago had to work if the city was to receive federal funds.	Net gain of newspaper support, because the largest newspapers opposed public housing for their own reasons (though some newspapers were campaigning strongly for the project). The interdependence of influences, and the net effects, are very uncertain here.

Some of the costs and gains as viewed by the mayor are similar in nature to those felt by the alderman. Yet the weights attached to them were no doubt quite different. Moreover, other effects on the mayor were quite different from those on the alderman. In both instances, however, the considerations obviously diverged greatly from costs and benefits from the community's standpoint, and the latter diverged sharply from costs and benefits from the nation's viewpoint. In general, the criterion of good policy used by each participant in the political process, though tempered and constrained by the influences of other participants, differs greatly from the criteria used in cost-benefit analyses.[14]

Transportation in Washington

Transportation proposals for Washington, D.C., provide another case that reminds us of the hurdles that planning must take into account. Although Washington, D.C. may not be a typical metropolitan area, it is a good subject of study in one respect: there are voluminous hearings on its problems conducted by Congressional committees. These hearings bring out rather vividly some of the divergences between costs and benefits as viewed by various officials and individuals, though here also many influences are felt by officials without any explicit testimony about those considerations being given.

Recent hearings pertained to a specific proposal to improve urban transportation in the Washington area.[15] Many citizens' groups opposed the use of freeways and cloverleafs because of the properties that would have to be taken over for rights of way, because of noise and other impacts on their neighborhoods, and because of beliefs that the equivalent job could be done less expensively by more effective use of existing streetcar and railroad tracks.[16] For some groups, however, there were opportunities for large monetary gains. It was alleged, incidentally, that General Motors, Firestone, Phillips Petroleum, and Standard Oil of California have financed National City Lines and that NCL buys control of local street railways with the understanding that they will substitute buses for streetcars.[17] (Sometimes firms are formed to buy up unprofitable transit companies and resell them at a big profit to a newly organized municipal authority.) Such activities, which supposedly run counter to the public interest, need not be illegal or immoral; they come about because cost-reward structures induce public officials to co-operate with these proposals when reaching decisions.

Another interest group comprised the existing local transit companies. Their main concern was survival, and they were worried primarily about the organization of the central transportation authority, because such authorities sometimes set up competing bus lines in such a way as to bankrupt the private companies.[18] Another interest was that of Lockheed, which proposed a monorail system.[19] Existing transit companies and many others took a dim view of this plan. Testimony of another group, the

[14]John Krutilla has pointed out that it is advisable to distinguish between (1) public outlays to make economically efficient investments, such as transportation aids or water-pollution controls, and (2) those avowedly intended to redistribute wealth. As Krutilla has stressed the appropriate cost-benefit measurements from a community standpoint depend upon the purposes of each proposal. This distinction is a relevant and an important one; though I believe that cost-reward structures, unless bargaining arrangements are just right, can distort either type of project--e.g., proposals conceived originally as ways of helping impoverished farmers as well as those intended as conventional investments.

[15]Transportation Plan for the National Capital Region, Hearings before the Joint Committee on Washington Metropolitan Problems, Congress of the United States, 86th Congress, 1st Session (Washington: U.S. Government Printing Office, 1960).

[16]Ibid., especially pp. 793-823, 873, 901-53, 981-89.

[17]Ibid., p. 983.

[18]Ibid., pp. 594-694.

[19]Ibid., pp. 605-606.

Bureau of Public Roads, seemed to be in favor of freeways. On one point all local groups seemed to be in agreement--namely, that the larger the share of the burden assumed by the federal government, the better the proposal (other things being equal).

The major costs and gains as seen by the adjacent states are easy to figure out. Representative Lankford of Maryland thought the plan (as modified after the first hearings) (1) did not allow for (encourage?) enough growth in portions of Maryland, (2) did not make efficient use of existing track and rail facilities, and (3) did not give the states enough voice in future transportation decisions.[20]

The point to be stressed here is that the divergent pulls exerted by cost-reward structures are omnipresent, not immoral or abnormal or exceptional. As we all know, sometimes cost-reward structures lead to graft and corruption.[21] But our concern here is not with violations of the law. It is usually true, as Blanshard wrote about New York City, that "The great majority of the city's employees are honest, industrious, and faithful. But the faithfulness of these employees rarely receives the headlines, whereas the exposure of one black sheep is featured in all the newspapers."[22] This still leaves the wrong impression, however, for it sounds as though corruption is the only thing that should cause us concern. But for every instance of corruption, there must be a hundred instances of distorted, inefficient, or completely frustrated programs produced by the cost-reward structures that confront municipal leaders.[23]

Cost-Gain Patterns

This bargaining process, which undoubtedly leads to more satisfactory results with some institutional arrangements than with others, is a pervasive phenomenon. "Even in totalitarian societies, where the opportunities for ordinary citizens to assert their demands and to organize to press for them without the sanction of the ruling cliques are tightly circumscribed, there are some evidences that this bargaining process occurs."[24] It certainly does occur in totalitarian societies--the differences between those societies and others are mainly in the amounts and kinds of influence possessed by various participants. Brute force, for instance, may be a major influence. In other words, the differences between institutional arrangements manifest themselves in the cost-gain structures that face various officials and individuals.

---

20 National Capital Transportation Act of 1960, Hearings before the Joint Committee on Washington Metropolitan Problems, Congress of the United States, 86th Congress, 2nd Session (Washington: U. S. Government Printing Office, 1960), p. 5.

21 See, for example, Paul Blanshard, Investigating City Government in the La Guardia Administration, A Report of the Activities of the Department of Investigation and Accounts, 1934-1937 [no publisher indicated], 1937, and James Reichley, The Art of Government: Reform and Organizational Politics in Philadelphia, A Report to the Fund for the Republic, reprinted from Greater Philadelphia Magazine, Philadelphia, Pa., [no date shown but probably appeared in 1959], pp. 1-38. Earlier, of course, there was the muckraking era during which Lincoln Steffens wrote his famous exposés.

22 Blanshard, op. cit. (footnote 21), p. 61.

23 For additional cases and discussions that are especially pertinent to these matters, see Edward C. Banfield, Political Influence (New York: Free Press, 1961).

24 Sayre and Kaufman, op. cit. (footnote 12), pp. 88-89.

Note how decisions change when cost-reward structures change. If the cost of something goes up, a business or individual consumer usually buys fewer units. If the cost of any action goes up, an individual is deterred from taking that action to a greater extent than before. If the worth of some item to a person goes up, he finds himself willing to pay more for it. If the value of some action rises, an individual is more anxious than before to take that action.

Often in politics, circumstances gradually alter the cost-reward structures confronting officials. Prior to the construction of the Idlewild airfield, the value to various groups of having more airport facilities gradually rose, and officials clearly felt the shifting cost-gain patterns from their own standpoints.[25] In Philadelphia, when support for two Republican leaders became almost evenly divided in 1956, the value of acquiring the support of a few more ward leaders soared, and "large bribes were reportedly offered to wavering ward leaders, and it is said that several of them were threatened with physical violence."[26] To mention another example, as the probability of success declined for the Republican party in Philadelphia, the expected value of contributing financial and other support fell also, and therefore the amount of support given to the party declined.[27] Shifts in cost-reward patterns are frequently introduced deliberately. When the federal government offers to pay 90 per cent of the cost of certain freeway projects, it makes the net gains to the local community almost irresistible. When Mayor La Guardia of New York began to get action in his campaign against corruption, he made certain illegal activities more costly than they had been. When costs and rewards change, decisions and actions change until a new temporary equilibrium is achieved. The bargaining process leads to a balancing of these forces somewhat as marbles seek a position when they are poured into a bowl, and further shifts in decision-making occur somewhat as the marbles alter positions whenever the bowl is tilted.

IMPLICATIONS FOR URBAN PLANNING AND COST-BENEFIT ANALYSIS

It is of growing importance for urban planners and cost-benefit analysts to keep these political realities in mind. We read repeatedly of the inability of planning agencies to accomplish much.[28] Would the existence of sound cost-benefit analyses really have affected the outcomes (without concomitant institutional changes)? Probably not. Many of us have seen fairly convincing cost-benefit or economic analyses have little or no impact on decisions. What should urban planners and cost-benefit analysts do?

One possible course of action is for them simply to lower their sights and be less dissatisfied with the imperfections of urban living. If our

---

25 See "Gotham in the Air Age" in Public Administration and Policy Development, Harold Stein, ed. (New York: Harcourt, Brace and Co., 1952), pp. 145-97.

26 Reichley, *op. cit.* (footnote 21), p. 14.

27 *Ibid.*, p. 18.

28 As an illustration, see William A. Robson, The Government and Misgovernment of London (London: George Allen & Unwin, 1939), pp. 186-91.

aspirations are too far ahead of our capacities, it is supposed to be healthy to relax a little. Besides, the present bargaining apparatus in American cities, even with all its shortcomings, does have virtues. It seeks to compromise among various pressures and allots at least some weight to most of the important interests. The outcome is not random, nor is it completely perverse. The results are better than we could expect from an extreme form of hierarchy.[29] If the urban planner simply lowers his sights, cost-benefit analysis should perhaps be used to compare marginal modifications of municipal policies or modest plans that have a reasonable chance of acceptance under present political arrangements.

To some extent, this sort of life adjustment is no doubt in order. But it is not very satisfying, because we know that it is sometimes possible to effect relatively large changes in urban conditions. We should not completely rule out more ambitious actions; the relevant alternatives surely include inducing conditions to adjust as well as adjusting ourselves to conditions. Moreover, there is a whole spectrum of bargaining arrangements, even in American cities, and some work better than others. At minimum, then, urban planners can reasonably seek (1) marginal modification of cities within present urban political frameworks and (2) changes in political frameworks that might produce improved bargaining processes.

In other words they can seek ways of manipulating the cost-reward patterns that confront various participants. One method is by seeking modified institutional arrangements, as political scientists and public administration experts have done for a long time. Occasionally these modifications have been discussed explicitly as ways of altering costs and rewards.[30] This is surely the fruitful way to consider planning proposals or institutional changes--for example, to ask what a strong-mayor form of government or a smaller Council does to the cost-gain structure that faces the various officials and groups. If the mayor can be made less dependent upon others, he may find the adoption of "rational" urban plans more rewarding but he may also find that being concerned about small minority groups is less rewarding than before. In any event, the way to analyze the effects of changes in the framework is in the light of impacts on cost-reward structures.

Another method of manipulating costs and rewards that needs cautious consideration is getting the federal government to use overt subsidies and penalties more frequently. These devices are extremely effective in shaping local cost-benefit patterns--they can make a freeway or a housing project or a pension plan exceedingly attractive to state or local governments. Moreover, extension of the use of federal subsidies appears to be politically feasible, for taxpayers across the country object surprisingly little to small payments per taxpayer for the benefit of particular areas or groups. But this technique is also a dangerous one to encourage. It sacrifices important advantages of the local bargaining process--the ability of small minority groups to exert at least a little influence, and the avoidance of extremes that the usual frustrating bargaining process insures. Having the central government shape local cost-reward patterns can be a powerful force for good--or for evil.

---

[29]Charles E. Lindblom, "Bargaining: The Hidden Hand in Government," RAND RM-1434-RC, The RAND Corporation, Santa Monica, Calif., 1955, pp. 1-44.

[30]Simon, Smithburg, and Thompson, op. cit. (footnote 2), pp. 451-87.

Still another way of influencing costs and rewards, though it is a weak one, is to develop and publicize good cost-benefit analyses pertaining to the principal alternative plans to be considered. If sound, such analyses can to some extent increase the costs to officials of pursuing "bad" policies.[31] If no one can see which action is better than another, in terms of community costs and benefits, then community interest is not directly and explicitly considered at all. If officials and voters can see a relatively sound and convincing case for one action, then it is at least slightly embarrassing (i.e., expensive) to ignore such evidence. One must admit, though, that this impact on cost-reward patterns is extremely weak. Since the general public as well as various political factions are hard put to decide whether or not an analysis is sound, no group can turn it into much of a weapon. Besides, very few persons are trying to maximize net community benefit.

There is another technique for influencing cost-reward patterns that deserves more exploration--the greater use, by urban governmental units themselves, of tax and compensation provisions to bring individual interests and the public interest closer into line. This is what we often attempt to do where there are wide discrepancies between private and total costs--e.g., tax the user of soft coal when it imposes heavy costs on others, subsidize or shelter inventors since their activity produces external benefits for which they are not compensated. Federal subsidies for freeways are attempts to bring local and national interests closer together, but, as mentioned before, extension of this technique poses serious problems. But urban areas themselves may be able to make greater use of this method. Advocates of urban planning measures might propose ways of charging more of the principal beneficiaries and compensating more of the principal losers instead of trying to sell a plan that distributes free windfalls to some officials and citizens and forces heavy costs upon others.

One way or another the major opposition has to be compensated when actions are taken in a voluntaristic society. At present the compensation takes the form of intricate horse-trading or log-rolling which produces many bad side-effects or reaches an impasse that blocks further action. I am suggesting that it is sometimes feasible to charge the gainers and pay off the losers overtly and that where feasible this method can produce more rational planning. If an urban plan produces net community gains, it should be possible in principle to compensate the opposition. In practice it should be possible to move a little further in this direction. Maybe we can rig costs and rewards so as to bring self-interests more nearly into harmony with community interest. Perhaps we can move a bit closer to institutions such "...that a decision which is (subjectively) rational from the standpoint of the deciding individual, will remain rational when reassessed from the standpoint of the group."[32]

To mention a simple example, the cost-reward structures facing aldermen and municipal officials can often be improved simply by raising their salaries. This change makes their official duties less costly in terms of sacrificing their business interests and makes it less urgent for

---

[31]John Krutilla has suggested two important points here: Such analyses can increase these costs to officials further if the analyses are (1) better insulated from the political process, freeing the studies from political judgments and fixing more clearly the responsibility of decision makers, and (2) more timely so that they can be brought to bear in time.

[32]Simon, op. cit. (footnote 7), p. 243.

them to have their official actions pay off personally. If the effects of cost-reward patterns are explained more vividly to the public (instead of letting them believe that moral decay accounts for most bad decisions), voters may be willing to incur the costs of higher salaries. Still more important, though, would be influencing officials' costs and rewards by affecting constituents' views. Here especially is where tax and compensation policies might be effective. People who will gain from a project will not be turned against it even if they must give up part of the gains. People who will lose are not as violently opposed to a project if they are at least partially compensated.

Already, of course, we charge for easily identifiable benefits (e.g., water for irrigation or power for households) and compensate for easily identifiable damages (such as buildings and sites destroyed or used for freeways). It is time to consider going further. Maybe we can charge a crude (and conservative) approximation of the benefits from control of floods or water-pollution. Perhaps we can award rough (and again conservative) compensations for expected declines in property-values due to rezoning or other changes. If such taxes and compensations were incorporated into planning proposals in a realistic fashion, more voters' interests would coincide with net community benefits, and as a consequence the interests of aldermen and city officials would also more nearly coincide with community interests.

To explore these possibilities of rigging costs and rewards, we would have to examine costs and benefits from the standpoint of major factions as well as from the community's standpoint. Estimates of such costs and benefits are not new. Opponents of legislation often present evidence to show how much they would be damaged. Newspapers sometimes point out the windfalls that would be bestowed on other groups. But we need to examine these costs and benefits to particular factions more systematically, to give them the attention and emphasis that they deserve. And we must try to perceive what kinds of pressures are felt by aldermen and other officials as a result of these costs and benefits and of the political framework.

It would be impossible, of course, to determine gains and costs of an urban plan from the viewpoint of each individual affected, but we might be able to see the major impacts on certain categories of persons. Past experience with rezoning, freeways, and control of water pollution gives some information about the gains in property values attributable to these actions. Past experience also tells us something about what groups would get hurt. Many such effects used to be measured and labelled "secondary benefits." If recognized as benefits (and costs) from the standpoint of particular groups rather than as net community benefits (and costs), such estimates might yet serve a valuable purpose.

We need these analyses of cost-reward patterns, it seems to me, whichever strategy urban planners or cost-benefit analysts adopt. We need to know more about costs and gains from various groups' viewpoints if we are to explore the extended use of charges and compensations to implement urban planning. We need this knowledge if we are to recognize what urban plans have a chance of being implemented under present political institutions--i.e., if we are to design cost-benefit analyses that compare politically feasible urban plans. We need such knowledge too if we are merely to understand the possibilities of urban planning better--if we are to understand the strengths, weaknesses, and dangers of alternative institutional arrangements in municipal planning.

8

Quality of Government Services

by Werner Z. Hirsch[1]

Interest in the quality of urban government services comes from a number of directions. First, demand and cost analysis requires that quality be made explicit. For example, either a cost or demand function of public education is not very useful where education is an unknown mixture of various shades of poor, mediocre and excellent qualities. A second interest relates to the need for performance comparison of different governments offering a given service. Government policy with respect to service quality is a third issue. Unlike price policy, quality policy is multi-dimensional and an urban government operating on a given budget should seek efficient ways to render maximum amounts of each of the various service qualities. Finally, there is interest in better understanding temporal quality changes, if for no other reason than to adjust production and price indices for quality changes.

An understanding of the quality aspects of government services requires consideration of what and how government produces and an examination of the uses to which its products are put. Government outputs are difficult to appraise and once estimates have been obtained, their meaning often cannot be readily interpreted. Let us illustrate these points by an example. Between 1902 and 1958, total current expenditures for local and state government services increased by 3,700 per cent. During the same period, similar expenditures for sanitation increased by 2,300 per cent, and expenditures for health and hospitals, 5,400 per cent. About the same relationship prevailed for capital outlays, with total capital expenditures for local and state government services increasing by 7,600 per cent, while those for sanitation went up 2,700 per cent, and those for health and

[1]The author is very much indebted to Professors Harold J. Barnett, Jesse Burkhead, and Jerry Miner; Drs. Selma Mushkin, Norman Breckner, and Norman Townshend-Zellner; Messrs. Elbert Segelhorst and Morton Marcus, for numerous helpful suggestions and criticism. Needless to say, all are accorded full discharge from responsibility.

hospitals, 20,700 per cent.[2] Do we have a series of output estimates? Do these figures imply that the service with the largest relative increase improved in quality more than the others? Specifically, does this mean that the quality of public health and hospital services during 1902-1958 improved more, relatively, than the quality of all government services? And were the quality increases of sanitation during this period relatively lower than those of other government services? These are important questions on which we hope to shed some light in this paper.

My plan is first to examine the work economists have done so far in defining and measuring quality of goods and services in general and of urban government services in particular. This appears appropriate since the quality issue has been somewhat neglected in the literature. I intend, next, to present an approach to the problem, which will then be examined in some detail with regard to two services--refuse collection and public education. Finally, the issue of measuring temporal quality changes will be considered.

QUALITY--THE ECONOMIST'S STEPCHILD

Economists in the past have shown little awareness or interest in quality as an economic parameter, in spite of the fact that Adam Smith already realized that "the time spent in different sorts of work will not always alone determine this proportion (between two quantities of labor). The different degrees of hardship endured, and the ingenuity exercised must likewise be taken into account." But Smith had also realized that "it is not easy to find any accurate measure either of hardship or ingenuity."[3]

Alfred Marshall recognized the existance of quality differences, but explicitly disregarded them, assuming "for the sake of simplicity, that all the corn in the market is of the same quality."[4] Simplicity was well served by this assumption! But in all fairness it must be stated that by disregarding qualities, great complications were avoided. Technical and economic conditions of bygone days rendered this simplification somewhat more legitimate than those of today's world. In the eras of Adam Smith and even of Alfred Marshall, product and service inputs were technologically simple and changed relatively slowly. Product differentiation was minor and government services were relatively few. Thus the operation of the market which then, more than ever since, conformed to the purely competitive model of Marshall, may have handled quality differences quite well by assigning corresponding prices to them.

Perhaps the first major attempts to define and measure quality more clearly were made by a group of agricultural economists in the late

---

[2] Special computations in co-operation with Data Classification and Research Branch of the U. S. Bureau of the Census.

[3] Adam Smith, The Wealth of Nations, Modern Library Edition (New York: Random House, 1937), Book 1, chap. 5, p. 31.

[4] Alfred Marshall, Principles of Economics, 8th Edition (London: Macmillan & Co., 1920), Book 5, chap. 2, p. 332.

nineteen twenties. They applied empirical methods in order to relate produce prices to quality.[5]

A few years thereafter, Edward H. Chamberlin took a careful look at the quality dimension and concluded that ". . . In the quality of the product itself [we may refer to] technical changes, a new design, or better materials; it may mean a new package or container; it may mean more prompt or courteous service, . . . a different location." He considered quality changes as resulting in distinctly different products, i.e., "'product' variations are in their essence qualitative, rather than quantitative; they cannot, therefore, be measured along an axis and displayed in a single diagram. Resort must be had, instead, to the somewhat clumsy expedient of imagining a series of diagrams, one for each variety of 'product.' "[6]

To Lawrence Abbott, quality has been "a multi-dimensional variable--a compound of numerous elements (e.g., in a necktie: size, shape, type of construction, color fastness), each of which is variable."[7]

In the early nineteen hundred fifties, concern with income elasticities of consumer goods led H. Theil and H. S. Houthakker to look into the quality problem.[8] Theil defined "a quality as a perfectly homogeneous good . . . [and] a commodity as a set of qualities."[9] He stipulated that a set of qualities is a commodity only if the prices have a certain functional relationship to each other during the period considered. Using this concept and finding that the prices of butter and oleomargarine are highly correlated, he looked upon them as a single commodity.

Houthakker introduced qualities as separate variables, which are determined by the consumer no less than is quantity.[10] Thus the consumption of the ith commodity is described by physical quantity X_i and quality V_i. The latter number indicates the quality bought and is defined as the price per unit under some basic price system. The total revenue from the sale of X_i units of quality V_i will then be $X_i V_i$. This is the simplified case, which is followed by one more complex in which the quality of a commodity is described by multiple quality variables. Thus, he stated, "If we are speaking of an overcoat we may, e.g., give quantitative expressions for its size, weight, colour, warmth, etc. The consumer will then be supposed to have preferences for various combinations of those characteristics, rather than for overcoats which are only described by their price. On the other hand, the price of the coat will also depend on these characteristics, since they are produced by different factor inputs."[11]

---

[5]Frederick V. Waugh, Quality as a Determinant of Vegetable Prices (New York: Columbia University Press, 1929); W. S. Kuhrt, "A Study of Farmer Elevator Operation in the Spring Wheat Area," Pt. 2, U. S. Department of Agriculture, Preliminary Reports, Washington, D. C., October, 1926, and October, 1927; Claude L. Benner and Harry S. Gabriel, "Marketing of Delaware Eggs," Delaware Agriculture Experiment Station Bulletin, No. 150, 1927.

[6]E. H. Chamberlin, The Theory of Monopolistic Competition, 5th Edition (Cambridge: Harvard University Press, 1946), p. 71 and pp. 78-79.

[7]Lawrence Abbott, "Vertical Equilibrium Under Pure Quality Competition," The American Economic Review (X/43), December, 1953, p. 827.

[8]H. Theil, "Qualities, Prices and Budget Enquiries," Review of Economic Studies, Vol. 19 (3), No. 50, 1952-53, pp. 129-47, and H. S. Houthakker, "Compensated Changes in Quantities and Qualities Consumed," Review of Economic Studies, Vol. 19 (3), No. 50, 1952-53, pp. 155-64.

[9]Theil, op. cit. (footnote 8), p. 129.

[10]Houthakker, op. cit. (footnote 8), p. 156.

[11]Ibid., p. 163.

The differentiation between the cost of creating quality and its value was earlier introduced by Hans Brems. He pointed to the two different meanings of the term quality as a parameter of action, i.e., "... what the consumer gets from the product ... [and] what the producer puts into it."[12] Brems thus advocated the use of consumers' and producers' criteria in defining and evaluating quality, surmising that these two coincide only in a few cases.

In the last few years emphasis on quality has come from a new source, i.e., from concern that production and price indices fail to reflect changes in quality adequately. In recognition of this issue efforts were initiated to develop a quality index that could be used to deflate production or price indices, principally by Richard Stone, Erland von Hofsten, Edward F. Denison, Richard and Nancy Ruggles, Frank de Leeuw, and Irma Adelman and Zvi Griliches.[13]

Very little research on the quality of public services has been carried out so far. In the 1930's, Clarence E. Ridley and Herbert A. Simon began a major investigation of measurement of municipal activities.[14]

While this study had a primarily conceptual orientation, another major effort, by the National Board of Fire Underwriters, has been mainly concerned with specific measurement in the field of fire protection. With the aid of the Board, a set of standards has been developed to provide an equitable basis for appraising the potential conflagration hazard of cities and for judging the adequacy of all phases of municipal fire protection.[15] These standards were arrived at with the aid of engineering principles and they take into consideration characteristics of the fire department, water supply, fire alarm system, police department, structural conditions in business districts, and enforcement ordinances relating to building construction and fire prevention. The final grading involves an engineering evaluation of the physical properties and manpower of a fire defense system from the standpoint of effective fire fighting and prevention. Grading has been carried out for most cities in the United States.

In the field of public primary and secondary education, a most ambitious quality measurement project has been initiated by the State of New York Education Department. Efforts are being made to develop workable school quality criteria, together with empirical studies to attempt actual measurements.[16]

---

[12]Hans Brems, Product Equilibrium Under Monopolistic Competition (Cambridge: Harvard University Press, 1951), p. 18.

[13]Richard Stone, Quantity and Price Indexes in National Accounts (Paris: Organization for European Economic Cooperation, 1956); Erland von Hofsten, Price Indexes and Quality Changes (Stockholm: Bokforlaget Forum, 1952), 135 pp.; Edward F. Denison, "Theoretical Aspects of Quality Change, Capital Consumption, and Net Capital Formation," Problems of Capital Formation (Princeton University Press, 1957), pp. 215-61; Richard Ruggles and Nancy D. Ruggles, "Prices, Costs, Demand, and Output in the United States 1947-57," in Relationship of Prices to Economic Stability and Growth, Joint Economic Committee, 85th Congress, Second Session, March 31, 1958, pp. 297-308; Frank de Leeuw, "The Measurement of Quality Changes," Proceedings of the Business and Economic Statistics Section, American Statistical Association, 1958, pp. 174-83; Irma Adelman and Zvi Griliches, "On an Index of Quality Change," Journal of the American Statistical Association (56), September, 1961, pp. 535-48.

[14]Clarence E. Ridley and Herbert A. Simon, Measuring Municipal Activities (Chicago International City Managers' Association, 1938).

[15]League of California Cities, The Fire Protection Grading Process as Related to the Economics of Fire Protection (Los Angeles: League of California Cities, April, 1961), p. 65.

[16]Samuel M. Goodman, The Assessment of School Quality (Albany: The University of the State of New York, 1959), 65 pp.; and William D. Firman et al., Procedures in School Quality Evaluation (Albany: The University of the State of New York, 1961).

In addition, certain benefit-cost analyses of urban government services that directly bear on this problem have been undertaken in recent years.[17] Furthermore the author has attempted to measure the quality of education, fire protection, police protection, refuse collection and street services.[18]

Henry D. Lytton, in a path-breaking study, has attempted to measure the output of certain federal government departments, including the Post Office, Veterans Administration and Internal Revenue Service.[19] Output measurements have mainly been made in terms of the number of such items handled as papers and letters. As the author recognizes, these measures tend to neglect the quality factor.

Henry J. Schmandt and G. Ross Stephens have used the number of municipal sub-functions performed as an indication of service quality level.[20] For example, they utilized data collected by the Citizen's Governmental Research Bureau of Milwaukee, which breaks down police protection into sixty-five categories, including such activities as foot and motorcycle patrol, criminal investigation, youth aid bureau, ambulance and pulmotor service, school crossing guards, radio communication, radar speed units, etc. They assume that the more sub-functions performed, the higher is the service quality. This approach has some serious shortcomings. For example, it takes into account neither the relative importance of a sub-function nor its quality.

All this work can provide a helpful point of departure for an analysis of the quality aspects of urban government services and their measurement.

SOME GENERAL CONSIDERATIONS

In our effort to gain better insight into the quality phenomenon and to attempt measurements, the following general outlook will prevail. Products of outputs of government, not unlike those of private enterprise, are considered to have both quantitative and qualitative aspects. The line is blurred for quite a few urban government services and in virtually all cases, identification and evaluation of quality characteristics is difficult. It is a key characteristic of most public as well as private services that

---

[17] Examples are Nathaniel Lichfield, Cost-Benefit Analysis in Urban Redevelopment (Berkeley: University of California, 1962), 52 pp.; Peter O. Steiner, "Choosing Among Alternative Public Investment in the Water Resource Field," American Economic Review, 49, December, 1959, pp. 893-916; Marion Clawson, Methods of Measuring the Demand for and Value of Outdoor Recreation (Washington: Resources For The Future, Inc., 1959), 36 pp.; John V. Krutilla and Otto Eckstein, Multiple Purpose River Development: Studies in Applied Economic Analysis (Baltimore: The Johns Hopkins Press, 1958), 301 pp.; Herbert Mohring, "Land Values and the Measurement of Highway Benefits," Journal of Political Economy, 69, June, 1961, pp. 236-49; Julius Margolis, "Secondary Benefits, External Economies, and the Justification of Public Investment," Review of Economics & Statistics, 39, August, 1957, pp. 284-91; Otto A. Davis and Andrew B. Whinston, "The Economics of Urban Renewal," Law and Contemporary Problems, 24, No. 1 (winter, 1961).

[18] Werner Z. Hirsch, "Determinants of Public Education Expenditures," National Tax Journal (13), March, 1960, pp. 29-40; and Measuring Factors Affecting Expenditure Levels for Local Government Services (St. Louis: Metropolitan St. Louis Survey, 1957) mimeographed.

[19] Henry D. Lytton, "Recent Productivity Trends in the Federal Government: An Exploratory Study," Review of Economics and Statistics (41), November, 1959, pp. 341-59.

[20] Henry J. Schmandt and G. Ross Stephens, "Measuring Municipal Output," National Tax Journal (13), December, 1960, pp. 369-75; and Harvey Shapiro, "Measuring Local Government Output: A Comment," National Tax Journal (14), December, 1961, pp. 394-97.

they do not conform to such clearly defined units of output as do wheat and corn, commonly used as examples by the classical economists.

Defining and measuring output units of government services does not necessarily offer unique difficulties, although they tend to be more pervasive than those of private goods. At the same time, defining social want-satisfying units appears more urgent than in the case of private goods, since the market mechanism that helps define private goods in money terms, and thus facilitates proper allocation, cannot be counted on with regard to public services.

Quality determination of urban government services can proceed in three distinct steps--

definition of service unit,

identification of quality characteristics of service unit, and

estimation of the money-value and money-cost, respectively, of quality characteristics.

These steps will now be examined in turn.

Efforts to define a basic urban government service unit cannot draw on tradition or usage. It becomes a matter of explicit analytical choice. The rendering of a service is an activity which has both demand and cost aspects, not unlike the production and distribution of a commodity. A government service, no less than any tangible commodity, can be viewed as having a variety of quality dimensions--both from the views of producer and consumer. As Brems has already intimated, there can be cases in which those who produce a service and those who consume it might want to argue in terms of different physical units. However, such a practice would at best be cumbersome and unworkable, since it would not permit internally consistent cost and demand analyses. Instead, it is essential that the same service unit be applied on both the cost and demand sides.

Defining a useful, basic service unit involves discovery of a high level of abstraction to which major qualities of concern to producers and users can be attached successfully. Where there is a choice of abstraction, the issue is one of value judgment. Thus, in our opinion, the basic government service unit should be defined in such a way as to be a unit of contribution to the successful pursuit of the aims of the government activity. In virtually all cases this will favor the consumer. But there is a further reason that induces us to prefer the demand side. Technological changes, including substitution of one type of service for another, can greatly modify the production process, yet the output continues to satisfy the same desires. For example, the evolution from a horse-drawn street car to a cable car, to a bus, and finally to a high speed electric train--perhaps a monorail--involves distinctly different processes which all, however, are designed to meet the urbanite's need to move rapidly and conveniently from one part of town to another. Even good highways, allowing private cars to travel fast and safely, are in a sense substitutes. Thus, to define the basic service unit as a certain technological process, or type of input, is not appropriate, since with the passage of time new processes will appear that will have different quality characteristics; they will usually defy comparison. However, regardless of improvements and changes, consumer preferences can often be judged by the same criteria involving virtually unchanged quality characteristics on the demand side.

Thus, the ideal basic service unit should be flexible and should accommodate the largest possible horizon of existing or potential quality dimensions. It should be defined in real terms and if a choice must be made between the demand and cost sides, the former should dominate.[21]

Identifying service units is a major undertaking and a few examples are in order. Let us start with residential refuse collection. The amount of refuse can be stated in pounds, cubic feet, or even number of containers. The refuse collection service provides a household (or person) with an average weekly pickup of a given number of pounds (or cubic feet or containers). Thus, refuse collection and disposal per household (or per capita) per week in pounds is a useful basic service unit.

In the case of water supply, the basic service unit is a cubic foot of water delivered to the place of use. The basic unit of street cleaning is a square foot of street cleaned, and that of street lighting, a mile of street lit.

For some urban government services, defining the basic service unit in terms of consumer preference is more complex. For hospital services, the unit might be a patient-day in the hospital; for police protection, a city block protected from crime; for fire protection, a city block protected from fire; and for schools, the education accruing to a pupil per day or year.

Once the service unit is defined, its relevant quality dimensions must be identified and ultimately measured. In these attempts it is helpful to remember that many government departments perform a service in a manner quite similar to a vertically integrated firm. In some cases the service unit involves a tangible product, which is produced by government and then delivered to users. Under such circumstances, the product itself can have quality dimensions, as can the delivery process.

Water supply is a good example. A cubic foot of water has important inherent quality characteristics in terms of its physical, chemical, and biological attributes. Among the major are hardness, turbidity, temperature, color, taste, odor, mineral content, bacteria count, etc. But a municipal water department involves a high degree of vertical integration in that it not only produces water, but also wholesales and retails it. There are various quality characteristics distinctive of water in the delivery process--water pressure, reliable supply, rapid repair, courteous and correct metering, etc.

A similar situation prevails with regard to library service, where a book is the basic unit. Selection and physical condition are important quality characteristics of the books themselves. Availability of books when requested, good reading room facilities, help to children in selecting books, reference service, and location of library, are all auxiliary quality characteristics of serving patrons.

In the case of refuse collection, discussed below, virtually all quality issues relate to the delivery process; the service is delivering residents from their refuse.

After defining relevant quality dimensions, measurements must be devised and two separate criteria are indicated. One should mainly

---

[21]The discourse of the service unit issue has benefited greatly from points made with great precision by Professor Richard A. Musgrave in his discussion of an earlier version of this paper and from discussions with Dr. Norman Townshend-Zellner.

reflect the hardship and ingenuity involved in rendering the service, already pointed to by Adam Smith. In short, we must measure the technical input factors, i.e., the cost side. The second measure should reflect consumer preferences and benefits, i.e., the demand side. For the sake of comparability, it is desirable that the quality characteristics of the service unit be translated into money-value on the demand side, and into money-cost, on the cost side.

An understanding of quality dimensions from the demand side, and the estimation of money-value, can be facilitated by identifying beneficiaries and types of benefits that are likely to accrue. For many urban government services, direct benefits are much more substantial and weighty than indirect. Unless our main concern is to facilitate efficient resource allocation, which requires equating marginal social benefits and costs and thus inclusion of as many indirect benefits as possible, the emphasis can be on direct benefits.[22]

Estimates of money-value of quality characteristics will usually require a benefit analysis. However, this is complicated by the fact that many quality characteristics are interdependent and interact in an intricate manner, a difficulty which also exists in relation to the money-cost determination.. Estimates of money-cost of quality characteristics can be made either with the aid of engineering cost functions or ex post empirical regression studies.

We will now turn to two examples, close to the ends of the spectrum of urban government services, and explore their quality aspects and measurements in some detail. Residential refuse collection has reasonably definable and measurable physical units and quality characteristics. The other example, public education, is perhaps the most intricate urban government service of them all, and many of its quality aspects presently defy measurement.

QUALITY MEASUREMENT OF RESIDENTIAL REFUSE COLLECTION

As previously noted, a useful basic service unit is refuse collection and disposal per household (or per capita) per week in pounds. However, such a unit can be associated with different service qualities. To determine significant quality dimensions, it is useful to ask who the possible beneficiaries of residential refuse collection are and how they are likely to benefit. In brief, it can be stated that refuse collection can affect the person at whose residence the pickup is made; those who live near the disposal site; those who are near streets used by pickup trucks; and the community at large, since refuse collection combats many health hazards. With regard to each type of beneficiary, various activities of varying quality can take place.

From the resident's view, the manner in which refuse removal is made, and especially its frequency, bear on the elimination of both odors and the breeding of insects and rodents; disease prevention; and cleanliness and beauty of the neighborhood. Quality differences are based mainly on the number of weekly pickups; care and reliability of the removal services;

[22]This point was eloquently made by Professor Richard A. Musgrave in formal discussion of an earlier version of this paper.

and cleanliness, noiselessness and courtesy of the collection crew. Quality differences also affect the effort the constituent must make, which is the convenience factor.

Thus, from the resident's view, the quality of residential refuse collection can be assessed mainly in the following terms:

number of weekly pickups

proximity of pickup location to building

nature of pickups, i.e., whether separation of refuse into garbage and trash is required

Based on these characteristics, eight different qualities of refuse collection can be identified. They are:

1. separate, once a week curb collection
2. combined, once a week curb collection
3. separate, twice a week curb collection
4. combined, twice a week curb collection
5. separate, once a week rear-of-house collection
6. combined, once a week rear-of-house collection
7. separate, twice a week rear-of-house collection
8. combined, twice a week rear-of-house collection

Similarly, refuse collection can be analyzed in relation to those who live near the disposal site. Modern incinerators will benefit this group more than city dumps or open land fills. Finally, transportation of refuse is less annoying and hazardous if it is made in specialized refuse collection equipment rather than in open trucks. Both items can be included in the specification of refuse collection quality, thus giving us five major quality dimensions.

Now let us turn to the measurement of some of these quality dimensions. On the cost side, each must be studied in terms of its technical factor inputs and their cost implications. While in theory many quality dimensions involve continuous data which can assume an infinitely large number of values, the alternatives are relatively few in the case of refuse collection service. For example, residential collection frequency is basically once or twice a week. Three collections are most unusual and represent the maximum. Likewise, pickup locations are either on the curb or behind the house. Pickups are primarily on a combined trash and garbage, or separate, basis. Thus, in this case, we mainly face sets of dichotomies. We would like to estimate the effect of either quality alternative on costs. Care must be exercised to isolate net relationships.

Let us look at an example. A field study of refuse collection and disposal operations in thirteen California cities made in 1950 and 1951 by the

Sanitary Engineering Research Project of the University of California, can help provide cost information on three quality dimensions, although the findings are mainly in terms of labor requirements.[23] It indicates that collection frequency is a major cost factor. For example, twice a week collection generally resulted in substantially greater quantities of refuse from each household each week than once a week collection. Specifically, the increase amounted to about 47 per cent. In addition, time studies of the effect of collection frequency on labor requirements of the pickup operation showed that twice a week collection required approximately 55 per cent more manpower per ton of refuse than once a week collection, assuming that about equal amounts are collected from each household each week. Therefore, twice a week collection may increase labor requirements by about 128 per cent. Since disposal and administration costs are affected to only a minor extent by collection frequency, and labor costs tend to comprise about 80 per cent of all costs, it might be possible to conclude that twice a week collection is about twice as costly as once a week. Once per week collection required an average of about 2.3 man-minutes per household per week and twice a week collection, an average of about 4.6 man-minutes.

Proximity of pickup location to building is a further quality dimension and there have been at least two empirical investigations into its cost. A study of the 1955-56 operations of a number of municipal and private refuse collectors in the St. Louis area revealed that in-street collection was about 35 per cent, and in-alley collection about 40 per cent, less expensive than rear-of-dwelling collection. The California field study indicated that the "approximate manpower requirements for pickup operation (time required to load the refuse on the collection vehicle) varied rather consistently from an average of 100 man-minutes per ton for 100 per cent alley or curb collection to 165 man-minutes per ton for 100 per cent, rear of house collection."[24] By making similar assumptions to those above, one might conclude from these two studies that rear of house collection may be about 50 per cent more costly than alley or curb collection.

No detailed empirical study known to us has addressed itself to measuring the cost differential of separate versus combined collection. But the California study looked into the cost of specialized collection equipment. It found that "the use of mechanical compaction type refuse vehicles requires approximately 10 per cent more manpower to pick up a unit of refuse than is required for open-body type trucks."[25] Such equipment is also more expensive and has fewer uses and may therefore be assumed to increase general costs by about 10 per cent.

If this information is applied to a simplified case where refuse collection has only two quality dimensions--collection frequency and pickup location--and the lowest quality case, once a week curb collection, is given the value 1, the following quality valuation, from the cost side, results--

refuse, once a week, curb collection	1.0
refuse, once a week, rear-of-house collection	1.5

---

[23]Sanitary Engineering Research Project, An Analysis of Refuse Collection and Sanitary Landfill Disposal (Richmond, California: University of California, 1952), 133 pp.

[24]Ibid., p. 2.

[25]Ibid., p. 2.

refuse, twice a week, curb collection 2.0

refuse, twice a week, rear-of-house collection 3.0

On the demand side, the money-value of given quality characteristics of weekly household refuse collection and disposal must be estimated. The benefits associated with frequency, proximity, and nature of the refuse collection service, as well as the disposal and hauling method, are mainly in terms of disease prevention, convenience and beautification. In the United States, only when collection and disposal practices are very poor can the health factor be assumed to play the major role. Assessing the value of convenience and beauty is a task we will not attempt.

In a few instances the market mechanism might help. There may be some communities in which households have a choice of buying other service qualities from private contractors, but even under such circumstances, price is unlikely to reflect quality differences significantly.[26]

QUALITY MEASUREMENT OF EDUCATION

In the area of public primary and secondary education, the relevant service unit is not entirely obvious. However, the education accruing to a pupil attending public school per day (or year) appears to be a useful measure. He is the direct beneficiary and although we also recognize that his present and future family, his employer, as well as society at large, are potential beneficiaries, we are tempted to neglect this issue here.

The education that accrues to a pupil in primary and secondary schools per day (or year) has a variety of quality dimensions. Kershaw and McKean[27] have grouped these dimensions as follows:

knowledge in standard subjects

knowledge in special and optional subjects

ability to reason

intellectual curiosity

"creativity"

social poise

emotional stability

physical health

---

[26]Some thought might be given to the possibility of using a game device, to permit a sample of refuse collection users to express preferences and indicate their value. The gaming could benefit from the experience Robert L. Wilson gained when he applied a simple game to the determination of service quality preferences in Greensboro and Durham, North Carolina. See Robert L. Wilson, "Livability of the City: Attitudes and Urban Development," Urban Growth Dynamics (New York: Wiley & Sons, 1962), pp. 387-99.

[27]J. A. Kershaw and R. N. McKean, Systems Analysis and Education (Santa Monica: The RAND Corporation, 1959), pp. 8-9.

Obviously these items are highly interrelated and often a school activity that affects one has a bearing on others. From this viewpoint, the money-cost (and money-value) of quality needs to be estimated jointly for all of these items. A more selective approach will be discussed later.

The quality of public education in general, and the acquisition of knowledge in standard subjects in particular, appear to be affected by many more factors than is the case in refuse collection. Empirical efforts toward their appraisal are further complicated by the fact that most of them are continuous variables. The following seem to be the most significant inputs:

1. Caliber of teaching staff and teaching load: Important characteristics are per cent of experienced teachers, per cent of teachers who are graduates of strong liberal arts colleges with majors in the field in which they teach; number of outside-the-area candidates interviewed for each teacher hired; number of students per counselor, librarian and other specialists; number of college hours the average teacher has taken; and per cent of teachers with more than ten years' teaching experience. Average teacher salary is likely to reflect some of these factors. Teaching load is another indicator. In many schools, twenty hours a week of teaching, and dealing with about 175 students, is considered normal. One measure would be the average teaching load of a given school in relation to the national average.
2. Caliber of school administration: The leadership and ability of the school superintendent and his principals cannot be neglected. Number of superintendents, principals and consultants per 100 pupils could prove a useful measure.
3. Grouping and class size: Many educators maintain that within limits good education requires students of common ability and interest to be grouped together.[28] The result is small classes, which generally indicate a more intensive education effort that can thus be measured by the pupil-teacher ratio.[29]
4. Teaching program: The scope and quality of the teaching program can be measured in terms of the number of high school instruction units offered by the school. Other measures might be per cent of college bound students who carry four courses a year in English, mathematics, science, history or foreign languages; per cent of students who take mathematics courses beyond a second year of algebra and one year of plane geometry, or four years of foreign languages.
5. Length of school year and day: Schools differ in the number of hours in a given year in which a child participates in formal education.

These characteristics combine in various ways and result in many different qualities of public education from the cost side. Basically, the estimation of the money-cost of the quality of education calls for determining the cost of various combinations of factors in the five main categories.

---

[28] See James B. Conant, The American High School Today (New York: McGraw-Hill, 1959), 140 pp.

[29] Like any other average, this ratio conceals much detail and should be separately given for primary and secondary education. Also the number of teacher aids should be considered.

For example, the cost implication of (a) the organizational structure of a school district, i.e., its use of an elementary and high school system (the 8-4 system) or the elementary, junior and senior schools (the 6-3-3 system), (b) its additional administrative and supplementary specialists, (c) its average class size, and (d) its average salary costs, can be readily appreciated from the sample district of Table 1. These ninety-six alternative per pupil school costs indicate the wide range of variability in quality associated only with select members of a district's professional staff.

The base quality represented in Table 1 is Case 1, in which the district employs an 8-4 system with two schools for its three thousand elementary grade pupils. Only the high school principal has an assistant and no professional librarian is employed. Students, in this case, are in relatively large classes--thirty-five and thirty students per teacher on the elementary and high school levels, respectively. The district which pays the lowest salaries (scale I) will spend \$99 per pupil annually for professional services, while one that is able to bid for somewhat better personnel (scale II), will have annual costs of \$146 per pupil; and one that pays salaries competitive with private enterprise for skilled administrators and with private academies and colleges for highly qualified teachers (scale III), will incur annual costs of \$248 per pupil. The highest quality represented in Table 1 is Case 32. Here the district uses a junior high school for improved student grouping, small elementary schools for improved teacher supervision, and smaller classes for closer pupil-teacher relationships. Principals in the six elementary schools have assistants and the junior and senior high school principals each have two. Each school has a trained librarian. The costs of this quality level are over 110 per cent higher than the base quality of Case 1 and there is a 432 per cent difference in costs between the lowest quality-lowest salary scale and the highest quality-highest salary scale.

In some cases, extreme paucity of data leaves the investigator either with the alternative of abandoning an attempt at quality estimation or of utilizing partial factor input information, which appears reasonably well correlated with overall service money-costs. Under such circumstances, the author used the number of principals, superintendents and consultants, per 1,000 pupils in average daily attendance, as a service quality measure for primary and secondary public education for selected years from 1900-1958.[30]

Next, the money-value of the quality of education will be considered. The pupil benefits from education depending to no small extent upon the earlier mentioned quality factors--caliber of teaching staff and school administration, grouping and class size, teaching program, and length of school year. Perhaps the single best way to estimate the money-value of different qualities of education is in terms of the resultant lifetime

[30] The use of this input measure assumes that the number of principals, superintendents and consultants contribute significantly to the school's service level, and that superior school districts not only hire more specialists than do inferior, but they are also of higher caliber. In general, it assumes that superior districts are rational and approximate a marginal calculus, i.e., the quantity and quality of all of the inputs that comprise good education satisfy the usual economic efficiency conditions, in which factors are employed to the point where the ratios of thei- marginal productivities equal the ratios of their prices. Werner Z. Hirsch, Analysis of the Rising Costs of Public Education, Study Paper #4 of the Joint Economic Committee of the Congress of the United States, 1959, 40 pp.

Table 1. Cost implications of alternative organizational plans, size of elementary schools, professional staff and classes, and salary scales for a hypothetical school district of 4,000 pupils

	Type of organi-zational plan		Number of Elem. Schools		Assistant Principals		Librarians per school		Teacher/pupil ratios		Cost per pupil ($)		
					Elem. 0	Elem. 1			Elem. 1/35	Elem. 1/20	Salary Scale*		
Case	8-4	6-3-3	2	6	Jr. & Sr. HS. 1	Jr. & Sr. HS. 2	0	1	Jr. & Sr. HS. 1/30	Jr. & Sr. HS. 1/15	I	II	III
1	x		x		x		x		x		99	146	248
2		x	x		x		x		x		104	154	258
3	x			x	x		x		x		104	153	259
4		x		x	x		x		x		109	161	272
5	x		x			x	x		x		102	151	255
6		x	x			x	x		x		108	160	268
7	x			x		x	x		x		111	163	274
8		x		x		x	x		x		117	173	290
9	x		x		x			x	x		102	150	254
10		x	x		x			x	x		108	160	267
11	x			x	x			x	x		110	162	273
12		x		x	x			x	x		116	172	289
13	x		x			x		x	x		105	155	261
14		x	x			x		x	x		112	166	276
15	x			x		x		x	x		117	172	289
16		x		x		x		x	x		124	183	307
17	x		x		x		x			x	178	262	444
18		x	x		x		x			x	191	282	475
19	x			x	x		x			x	183	269	455
20		x		x	x		x			x	196	289	489
21	x		x			x	x			x	181	266	451
22		x	x			x	x			x	195	288	485
23	x			x		x	x			x	189	278	471
24		x		x		x	x			x	203	301	507
25	x		x		x			x		x	181	266	451
26		x	x		x			x		x	184	288	484
27	x			x	x			x		x	189	278	470
28		x		x	x			x		x	203	300	506
29	x		x			x		x		x	184	270	457
30		x	x			x		x		x	198	293	493
31	x			x		x		x		x	196	287	485
32		x		x		x		x		x	210	311	526

*Position	Salary scale ($ per year) I	II	III
Principal, High School	5,800	9,000	17,000
Principal, Jr. High School	5,100	8,000	14,000
Principal, Elementary School	4,900	7,000	11,000
Assistant Principal, High School	4,500	6,500	10,000
Assistant Principal, Junior High School	4,000	6,000	9,200
Assistant Principal, Elementary School	3,700	5,500	8,700
Librarian, High School	4,200	6,000	9,200
Librarian, Junior High School	3,800	5,500	8,500
Librarian, Elementary School	3,500	5,000	8,100
Teacher, High School	3,600	5,100	9,000
Teacher, Junior High School	3,200	4,800	8,100
Teacher, Elementary School	3,000	4,500	7,500

earnings. While estimates of the effects of elementary high school and college education on lifetime earnings have been made, we are not aware of studies that have attempted to relate the earlier discussed qualities to lifetime earnings.[31] Such studies appear feasible and should be undertaken, keeping in mind that not all earnings increases are likely to have resulted from quality variation, but are also likely to be associated with the pupil's inherent ability and informal education in the home. For this reason it appears desirable to adjust earning results according to at least three factors in addition to educational quality--native ability or inherent intelligence, home environment, and motivation or ambition to learn.[32]

An evaluation of quality dimensions of public education can also proceed on a more restricted basis. It can be in non-monetary terms, which can then ultimately be translated into money-value terms.

Of the eight principal dimensions of education referred to earlier, there appears to be general agreement that a most important, perhaps the most important single item, is the acquisition of knowledge in standard subjects. Educational achievement in the basic subjects is measured with the aid of achievement tests. While the scores on batteries of standardized achievement tests are far from perfect indicators of output (they do not cover all of the worthwhile aspects of learning in the subject area), nevertheless it is possible that a comprehensive battery, like the Iowa Test, is sufficiently sensitive to produce a fairly good measure of the average academic achievement. Thus it is possible to use the scores of achievement tests, taken in senior high school, to reflect, in a major degree, quality dimensions of primary and secondary education, from the demand side.

At present, unfortunately, various school systems use different achievements tests and results are not directly comparable. Under these circumstances some makeshift methods could be employed to compare test scores. For example, the national average of test scores of a number of different achievement tests could be used as a norm. The corresponding figure of a particular school district could be compared to this norm. Thus, for example, while 30 per cent of all students who took achievement tests had a test score of a given value, different percentage figures would prevail for specific school districts. The larger the percentage above the norm, the higher would be the quality level of the school district, from the pupil's viewpoint.

Since a pupil's performance is not only the result of the quality of education to which he has been exposed, adjustments must be made similar to those suggested with respect to the lifetime earnings measure, i.e., native ability, home environment, and motivation.

---

[31]On the relation between educational attainment and earnings, see: G. Becker, "Underinvestment in College Education?" American Economic Review, Papers and Proceedings, Vol. 50, May, 1960, pp. 346-54; H. S. Houthakker, "Education and Income," Review of Economics and Statistics, Vol. 41, February, 1959, pp. 24-27; H. P. Miller, "Annual and Lifetime Income in Relation to Education: 1939-1959," American Economic Review, Vol. 50, December, 1960, pp. 962-86; E. F. Renshaw, "Estimating the Returns to Education," Review of Economics and Statistics, Vol. 42, August, 1960, pp. 318-24.

[32]No fool-proof information is available for these factors. However, the first factor might be approximated by IQ test scores, and the second, by the number of books read by the parents in a year. Adjustment for these factors can be made with the aid of co-variance or regression analysis. Martin David, Harvey Brazer, James Morgan and Wilbur Cohn, Educational Achievement--Its Causes and Effects (Ann Arbor: Survey Research Center, 1961), 158 pp.

A study along some of these general lines was recently completed by James H. Crandall.[33] It covered 6,000 children from grades 4 through 8 in sixteen California elementary school districts. Approximately one-half of these children were from eight school districts included in the top 10 per cent based on expenditures for instruction for all elementary school districts of 1,000 to 4,000 average daily attendance in California, over a 4-year period. The other half of the group were from eight school districts of similar size, but included in the bottom 10 per cent in expenditures for instruction during the same period. Each individual child had remained in the same school district for the full 4 years and had taken the same standardized intelligence and achievement tests. To avoid comparison of variations in achievement possibly resulting from differences in the intelligence of the children, all achievement test scores were considered in terms of their IQ. Expenditures on eleven items were related to academic achievement--administration, supervisors, principals, teachers, librarians, school clerks, instructional supplies, textbooks, all visual aids, workbooks, and health services. Significantly, higher academic achievement was separately associated with higher expenditures, both in terms of dollars and per cent of total expenditures for instruction, in each of the following categories--administration, principals, instructional supplies, workbooks, clerks, and health services. The study found that in all areas of academic achievement, districts high in expenditure for administration ranged from three to nine months above those school systems low in expenditure in this classification. These differences in average achievement were, in general, statistically significant at the 5 per cent level. They occurred between school districts which on the average, over a four-year period, spent $18.12 and $9.48, respectively, per pupil for administration. An even more consistently significant relationship of all areas examined occurred between academic achievement and expenditure rate for principals. The top four districts in this class of expenditure averaged both six months higher in academic achievement, and $10.30 more per pupil, than did school systems which spent the least money for principals.[34]

While achievement test scores tend to measure quality in non-monetary terms, it should be possible to relate a pupil's test scores to lifetime earnings after adjusting for the other factors that can affect his earnings, as was discussed above.

MEASURING TEMPORAL QUALITY CHANGES

Only measures which pertain to a single time period have been discussed so far, with one minor exception. Often, however, temporal comparisons are required. For example, it would be very helpful to initiate a quality series to adjust the U.S. Department of Commerce's implicit deflators for the state and local government sector of gross national product. Thus, Richard and Nancy D. Ruggles are convinced that the implicit price deflators do not properly allow for quality and efficiency improvements.

---

[33] James H. Crandall, "A Study of Academic Achievement and Expenditures for Instruction" (unpublished doctoral dissertation, University of California, Berkeley, 1961), pp. 1-33.

[34] This study validates to some degree the appropriateness of selecting number of principals, superintendents, and consultants, per 1,000 pupils in ADA, as an important quality measure, as was done in Hirsch, Analysis of the Rising Cost of Public Education, op. cit. (footnote 30).

They note that "the price of Government services as measured by the pay of Government employees" rose by an average of over 5 per cent a year from 1946 to 1957 and add, "there is a good reason to believe, however, that the productivity of Government workers has increased substantially in this period."[35]

In theory, we could use any of the measures that were discussed in the earlier section, make estimates for successive years, adjust them for price level changes, and so obtain estimates of temporal service level changes. Paucity of data appears to make such a step difficult, if not impossible. There are two additional methods for measuring temporal changes in service quality, as viewed from the cost side.

An effort could be made, for example, to decompose an expenditure series of an urban government service into two major components--polygenetic cost changes (those that are independent of service quality changes) and cost changes that result from changed service quality. Emphasis should rest upon obtaining good estimates of polygenetic cost changes. This figure should be subtracted from total cost changes and the residual would indicate service quality related cost changes.[36]

This method appears to offer little advantage, however, for services whose inputs are mainly in the form of labor or have greatly benefited from capital improvements during the period under analysis. It might, for example, prove useful for the assessment of service quality changes of municipal water departments and sewage disposal systems. For instance, chemicals and electricity are the major current costs of a municipal water department and neither element has undergone major quality change during the last few years.

Most urban government services, however, mainly use labor as input. For such services, a method might be developed that relies principally on wage increase comparisons. The assumption could be made that the government unit that pays higher than competitive wages and salaries is likely to procure higher than average quality labor, and that high quality labor also produces high quality output. Thus, if unit costs of a particular government unit have not changed over time, or have changed less than wages, and yet wage rates have increased more than in an industry which actively competes for the labor, the service quality might be expected to have improved over the competing industry's.[37]

Such a method, for example, might be applied to public and private education. On the average, whichever of the two pays better salaries is likely to acquire better teachers, superintendents, principals, and consultants. Assuming the same ratio of teachers, superintendents, etc., to students, the system that pays higher wages is likely to offer a higher quality of education.

In summary, it is apparent that efforts to measure the quality of urban government services offer exciting challenges and prospects. Much more work is needed along three key lines of inquiry--defining service units in real terms, identifying their major quality characteristics, and estimating the money-value and money-cost of these characteristics.

---

[35] Ruggles and Ruggles, op. cit. (footnote 13), p. 299.

[36] For a general discussion of this approach to measuring quality changes of private goods, see Adelman and Griliches, op. cit. (footnote 13).

[37] It would be useful to have information on quality changes in the industry which competes for labor. This industry could be producing a private good and such estimates might be possible. On this basis, the service quality change could be further tied to the quality change in the private good.

9

Spatial and Locational Aspects of Local Government Expenditures

by Seymour Sacks[1]

In this paper the spatial and locational pattern of local government expenditures in urban areas is analyzed with the use of suitable criteria of urbanization. Conclusions are drawn concerning local decision-making procedures as they are reflected in the expenditure patterns of the Cleveland area and several New York State urban areas. Existing analyses have emphasized per capita measures; this paper supplements them by using measures of expenditures per square mile which are directly comparable to quantitative measures of urbanization. While the initial utilization of the per square mile expenditures measure is restricted to nonschool expenditures, the results indicate that there are clear-cut spatial relations and that these measures may in fact explain to some degree the variations in per capita expenditures. The areal measures used also suggest that it is possible to distinguish urban and nonurban levels of expenditures which are indistinguishable when they are expressed in per capita terms. The results are thus tentative and restricted to those classes of expenditures most closely related to urbanization. Further, expenditures which involve special responsibility, such as welfare hospitals, and utilities, are excluded from the first stage of analysis.

Spatial considerations as used here refer to the position of a given community with respect to the total urban area under consideration; i.e., with regard to its placement, either as a central city, inner core, or outer ring community. Locational considerations emphasize the effect of a given community on its neighbors, the effects of proximity and contiguity on expenditure decisions.

While this paper is designed to emphasize methodological consideration in dealing with local governments in urban areas, it is hoped that the

[1]I should like to thank John J. Carroll, formerly Chief of Municipal Research, New York State Department of Audit and Control, for his many suggestions, comments, and enthusiasm which have been incorporated into this paper. Also, I should like to thank Alan K. Campbell, Jesse Burkhead, and Harold Pellish for their suggestions and comments.

specific conclusions drawn concerning local finances in the Cleveland, Syracuse, and Rochester urban areas can serve as hypotheses for the analysis of governments in other urban areas throughout the United States. In this analysis, as in almost all other analyses of local government expenditures, no attempt has been made to analyze the intra-jurisdictional distribution of local government expenditures. By phrasing the analysis in spatial and locational terms, however, this very important intra-jurisdictional problem is brought sharply into focus. A very much more advanced analysis than is contemplated here is necessary to reach meaningful answers concerning intra-jurisdictional variations.[2]

RELATING EXPENDITURES TO MEASURES OF URBANISM

Analyses of the factors determining local government expenditures have heretofore emphasized cross-sectional, and, to a lesser extent, inter-temporal comparisons not immediately related to urbanism. In the process of determining the links between existing analyses of local government expenditures and analyses of urbanism, a fundamental lack of coordination is apparent. Definitions of the urban area are either too broad, as in the Standard Metropolitan Statistical Area concept, or too vague from a governmental point of view as in the Urbanized Area concept. Existing methods of metropolitan and urban analysis are thus inappropriately grouping together local governments.[3] Only in a few instances has the unit of analysis been appropriate for the problem. In most cases the discussions have either considered a class of government, primarily cities, or have ignored the governmental unit entirely. In addition, the methods of dealing with government expenditures have been considered without due regard to the distribution of resources and functional responsibilities.

In general, the definitions that serve as a basis for establishing urban areas include both too much and too little--too much area and too few governments. With the exception of a small number of very populous counties, urbanization cannot be measured in units as large as the county.[4] Further, since consideration is usually restricted to cities above a certain size, villages and unincorporated areas are excluded from consideration as urban. Also excluded from consideration are school districts and other special districts which may be urban, but for which data are not available. Thus, the fragmented and overlying layers of government characteristic of most urban areas are also ignored.

---

[2]On the other hand, analyses of intra-jurisdictional variations in local government expenditures such as those carried on by William L. C. Wheaton and Morton J. Schussheim, The Cost of Municipal Services in Residential Areas (Washington: U. S. Housing and Home Finance Agency, 1955), and Walter Isard and Robert E. Coughlin, Municipal Costs and Revenues Resulting from Community Growth (Wellesley, Mass.: Chandler-Davis Publishing Co., 1957), are not phrased in terms suitable to local government fiscal problems. Moreover, they emphasize capital considerations which are entirely different in order of magnitude for any given period of time than are the operating expenditures which are the prime considerations of this paper.

[3]See Donald J. Bogue (ed.), Needed Urban and Metropolitan Research, Scripps Foundation Studies in Population Distribution, No. 7 (Miami Univ.: Scripps Foundation, 1953).

[4]Only twenty-five counties (including five city-county units) had population densities in excess of 1,500 per square mile and seven of these were those in the New York Standard Consolidated Area, and three in the Philadelphia area. These figures are based on the analysis presented in Local Government Finances and Employment in Relation to Population: 1957, U. S. Bureau of the Census, State and Local Government Special Studies No. 45 (Washington: Governments Division, Bureau of the Census, 1961).

A quantitative urban concept which is co-ordinate with government boundaries is not only essential to the analysis of local government expenditures and the local decision-making process but is of very great value where the urban concept has become intertwined with metropolitanism. In dealing with an empirical problem, such as the spatial and locational aspects of local government expenditures in urban areas, the model chosen influences the results to a very high degree. The same is true of other uses of the urban concept. Establishment of a model that permits a quantitative and comprehensive analysis of the fragmented and overlying layers characteristic of local government in general, and particularly of local government in urban areas, is therefore of fundamental importance.

Intensive application of existing criteria to New York State gave evidence of the fundamental lack of co-ordination mentioned earlier. By modifying the "Urbanized Area" concept for New York State and through the use of townships and cities as the basic units in delimiting urban areas a far greater degree of co-ordination with available data is achieved.[5] However, since units smaller than the county are employed, measures other than those based on population density are necessary. Property valuation densities for the various governmental units furnish an operational means for utilizing the Urbanized Area concept. This new definition of the urban area is co-ordinate with political boundaries and data.

The above considerations have demonstrated that while the urban problem is of great magnitude and complexity, it is probably more restricted in area and hence more manageable in extent than is commonly assumed. There is no doubt that the bulk of local government expenditures is made in urban areas. This is even more true when locally financed expenditures, primarily those which are tax supported, are considered. On the other hand, per capita expenditures in cities and other urban areas no longer are higher than in rural areas, at least in New York State. Per capita expenditures are in fact larger in many rural areas, and the same is true of expenditures measured per unit of property or per unit of income.[6] This has been the result of the increasing importance of highway and school expenditures and the lower proportion of children going to publicly supported schools in urban areas. State aid is the critical factor as the emphasis shifts from urban to many rural areas.

Within urban areas there has been a redistribution of functional responsibility as the dominant central city has declined in relative importance. Further, as noted before, urbanism has moved beyond the "walls" of the city and has assumed a new form in many places. These new developments have had spatial and locational characteristics whose influence on local government expenditures are the subject of this paper.

---

[5]The use of townships for other states, in which such units exist, would considerably enhance accuracy. (See Vol. 1, p. 20, U. S. Bureau of the Census, Census of Government, 1957.) They exist in many of the states which have urban areas; not only in New York, but Ohio, Illinois, Indiana, Wisconsin, Michigan, which are prominently listed in the list of dense counties and which contain urban areas whose boundaries could be more accurately analyzed.

[6]See Comparison of Revenues, Expenditures, and Debt: 1949-59, Comptroller's Studies in Local Finance, No. 1 (Albany: Department of Audit and Control, 1961), pp. 25-26. This document contains an analysis of the components that make up the county over-all totals.

DEFINING URBAN AREAS

The problem of defining urban areas as entities apart from urban places is a relatively recent phenomenon. Sociologists, geographers, anthropologists, political scientists, city planners, philosophers, and others in addition to economists have probed the question as to what is urban. The results of a survey of the literature are not too encouraging.[7] Apart from the Census, terms are used rather vaguely and measures which must by their nature serve a multiplicity of purposes are accepted without further analysis. There is a tendency to overlook the political implications of the measure offered. While most scholars now have recognized that the term "city" is no longer synonymous with urbanism, they have not gone beyond the negative aspect of the problem. Apart from definitions, cities can contain nonurban as well as urban portions, e.g., many of the cities in Texas and Oklahoma include large nonurbanized areas. And cities certainly differ in their urbanization. Unincorporated areas may be more urban than cities. For instance, in the New York Metropolitan Area, especially in Nassau and Westchester counties, villages may be larger and more urban than cities. The village of Scarsdale, with its almost $2.4 million municipal budget for 6.4 square miles, is certainly not rural. However, most villages in New York State have rural, rustic characteristics. Arbitrary measures based on population size and legal designation are not adequate in dealing with urbanism.

New York State

In analyzing local government expenditures in New York State we are confronted with the problem of differentiation between urban and nonurban areas, especially where different kinds of local governments are involved. Some cities in the state do not have the size or the density of some villages, or for that matter of some entire unincorporated town areas; e.g., unincorporated portions of towns in the New York Metropolitan Area have higher densities than many cities and villages. Population densities based on the figures in Local Government Finances and Employment in Relation to Population: 1957[8] show Nassau County exceeded by only eleven counties in the entire nation, all with at least one major city.[9] The largest community in Nassau is the village of Valley Stream with 38,000 people. On the other hand, some New York cities, such as Saratoga, Rome and Oneida, contain within their geographic boundaries characteristics which make them resemble rural areas rather than urban areas. In analyzing data on cities and villages as legally different classes the separation is relevant but when one considers them as urban local governments it is not.

---

[7]Robert E. Dickinson places the date of Christaller's pioneer work, Die zentralen Orte Süddeutschlands, dealing with the relationship of the town to the surrounding area in the early 1930's. Previous work dealing with the individual town and its functions, unrelated as they are to the surrounding areas, are inappropriate for an analysis of urbanism. The literature is still primarily focused on the towns as such, rather than on the urban area. The same is even more true of work derived from Park and Burgess which also tended to emphasize parts of the city rather than the urban area as a whole. "The Scope and Status of Urban Geography," Readings in Urban Geography, Harold M. Mayer and Clyde F. Kohn, eds. (Chicago: University of Chicago Press, 1959), p. 20.

[8]Op. cit. (footnote 4).

[9]Only Arlington County, Virginia, has a density which comes close to that of Nassau County without having a major city within its confines.

In the New York analysis the urban area concept is modified in terms of the political units involved. But there are other ways of identifying urban phenomena apart from population density and the special conditions enunciated by the Bureau of the Census. The special conditions for not excluding areas from the urban classification indicate that they involve very intensive land use or explicitly nonintensive recreational uses, both of which are characteristic of urban areas. For individual governments, a measure which encompasses residential as well as nonresidential land uses is the full valuation of real property per square mile, i.e., real estate and land.[10]

When criteria of taxable values are used, the phenomenon of urbanism falls into the correct perspective. Overlapping between rural and urban land uses and between the various intensities of urban land use is eliminated. The problems involving assessment data remain, but in broad comparisons these are of the second order of significance. Even interstate differences in what is included in the tax base are of relatively minor importance. One does not find changes in the urban nature of the community at the political border when the population size or density decreases if intensive land use for nonresidential, but highly valuable uses, is encountered.

To delimit the urban community, as well as to give the concept of the urban area a quantitative dimension, recourse is made to property valuation densities per square mile. The measures of urbanization thus constituted are thus co-ordinate with governmental jurisdictions. The urbanized areas are recognizable by the peaks in valuation per square mile and by the fact that there are fundamental differences between the isolated jurisdiction and one which is part of an entire urban area.[11] In New York State an arbitrary figure of $1 million of fully taxable value per square mile has been chosen because it conforms with the Census' Urbanized Area concept on a jurisdictional basis.

The emerging distinctions are clear cut. At the lowest end of what appears as a rural-urban continuum are the towns (akin to the New England townships) with equalized real property values of less than $20,000 per square mile or about $30 per acre. A rough guess that the true value was about 25 per cent greater than the equalized value would still give an acreage figure of $37.50. The next class involved values between $20,000 and $200,000 per square mile. Most of New York State's area falls in this category. Included in this class were several towns in counties whose total areas are considered as metropolitan.

The next category involved governments which had valuations between $200,000 and $1,000,000 per square mile. This category involved isolated jurisdictions as well as jurisdictions that fell into the urbanized fringe area. Included in this category are several cities that have large unurbanized portions which are reflected in their low property densities per square mile.

---

[10] In New York, full valuation of real property for each government (city, town, village, school district, and county) is determined by the Bureau of Equalization and Assessment of the State Office of Local Government. This measure is related to, but it is not equal to, the market value.

[11] See John J. Carroll and Seymour Sacks, "The Property Tax Base and the Pattern of Local Government Expenditures: The Influence of Industry," The Regional Science Association, Papers and Proceedings; Vol. VIII, 1962.

The next category between $1 million and $10 million involves the transition to the truly urbanized jurisdiction. Values of $1 million appear as transitional in the rural-urban continuum. Isolated communities may fall into the lower portion of this $1 million to $10 million category, but they involve no large land areas. Analysis of several states in addition to New York indicates that such presumed special cases as those jurisdictions which contain oil wells, iron ore pits, etc., would not meet the value qualifications except where the jurisdiction in question is very small in area, i.e., less than ten square miles. Urban land and real estate values including those on the fringe are of an entirely different order than comparable value in nonurban uses.

Above $10 million per square mile we find the central cities of metropolitan areas, as well as highly developed surburban communities. At its extreme, on Manhattan Island, the value of property exceeds $440 million per square mile of fully taxable property, i.e., excluding exempt property.

Figure 1 reflects the proposition that high density of population per square mile will yield a high property density per square mile, but that a high property density will not always reflect the converse. The measure permits the incorporation of the daytime population as well as the nighttime population in the definition of the urban area in units which are coordinate with the decision-making units of government.

On the basis of the property densities only the New York Metropolitan Area can be said to be unambiguously urban. No entire county in New York State outside of the New York Metropolitan Area can be said to be urbanized and in fact small proportions area-wise are involved. Further, there is no relationship between the per capita full valuations and the property values per square mile. Urban areas do not show the highest values per capita, those are usually special isolated cases involving a seasonally-used property, utilities, or state forest lands. On the other hand no isolated village no matter how small its area and no matter how large its bonanza can rival the urban areas--no matter how measured--in their property valuation densities per square mile.

The measure based on town and city property densities is very close to the "Urbanized Area" concept, but it has the advantage of being related to political decision-making units. As shown in Table 1 while the Standard Metropolitan Statistical Area implies a metropolitan and hence some urban characteristics to an entire county, the Urbanized Area concept does not. Using the property valuation densities for entire cities and towns does not show any marked difference in the proportion of a county's population called urbanized, as compared to the urbanized area concept.[12]

As noted earlier, a far smaller proportion of the area is in fact urbanized than is implied by the Standard Metropolitan Statistical Area concept. At its extreme in Oswego, Madison, Herkimer, Saratoga, and Rensselaer only the barest fraction of these counties' areas and a small fraction of the population are involved even by the most liberal definition of urban.

---

[12] Only in the case of Rockland and Suffolk counties is there a major difference between the two concepts. The use of a $2 million figure per square mile would tend to iron out the discrepancies. However, the higher values of land in these two counties reflect their transitory nature which have not yet been reflected in their population densities.

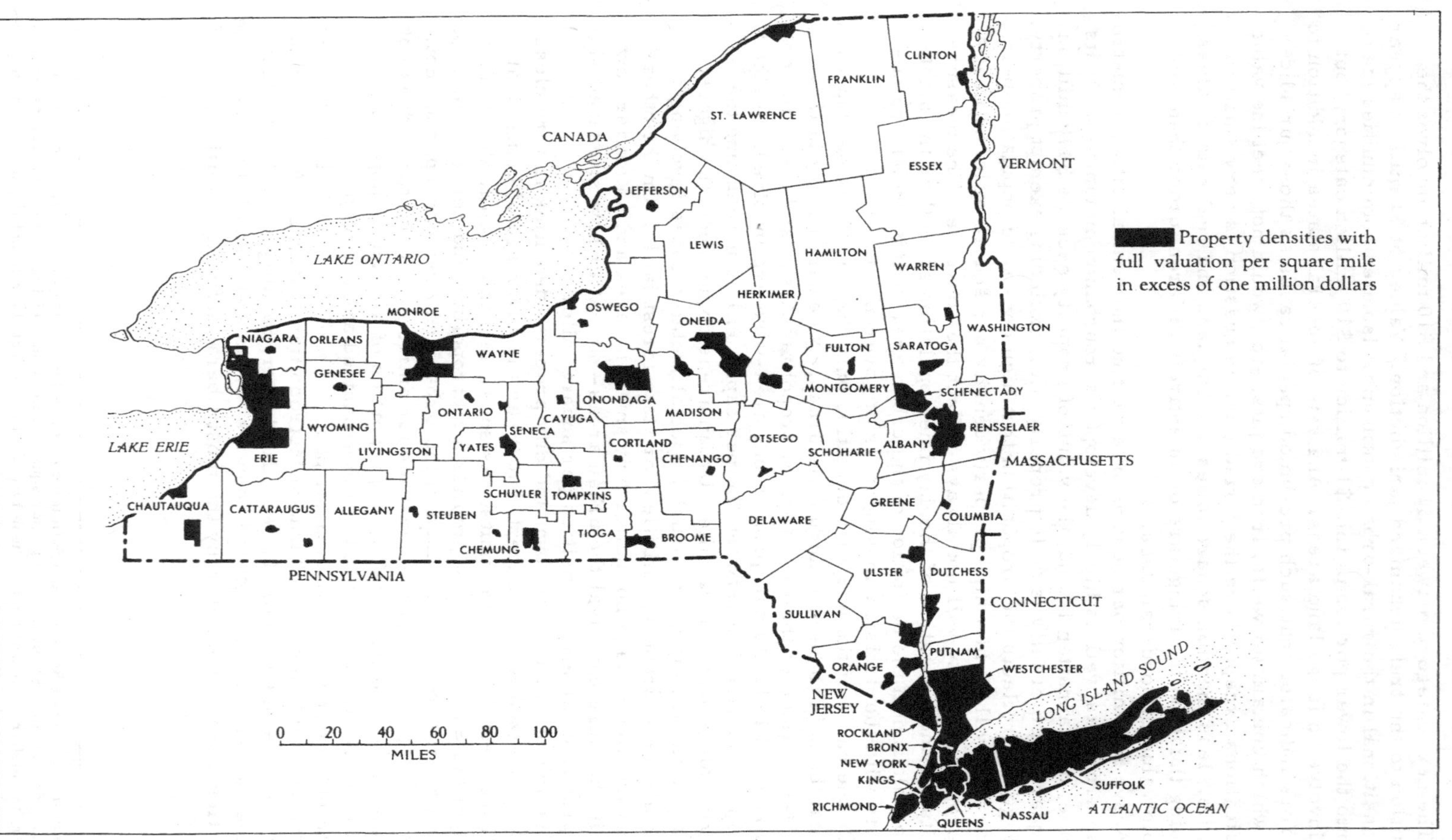

Figure 1. Urbanization in New York State, 1959.
Source: Comparison of Revenues, Expenditures, and Debt, 1949-1959, Comptroller's Studies in Local Finance, Number One (Albany: New York State Department of Audit and Control, 1961).

Table 1. Alternative measures of urbanization - New York state, 1959-1960

County	Standard metropolitan Statistical Area*		Classification based on density of valuation in cities and towns, by county**		
	Per cent metro-politan	Percent population in urbanized area	Percent population in urban area	Percent area urbanized	Density of valuation ($ million)
New York SMSA					
New York City. .	100	100	100	100	76.8
Nassau	100	99.7	100	100	16.0
Rockland.	100	63.7	100	100	2.0
Suffolk	100	48.9	100	100	2.5
Westchester . . .	100	92.2	99.7	95	8.9
Binghamton SMSA					
Broome	100	74.4	69.1	7	0.9
Buffalo SMSA					
Erie.	100	84.8	91.6	34	3.5
Niagara	100	63.2	68.6	19	1.6
Albany SMSA					
Albany	100	83.5	87.6	25	1.9
Rensselaer	100	61.8	66.8	8	0.5
Saratoga	100	8.0	8.1	8	0.3
Schenectady. . . .	100	86.4	97.4	54	2.9
Rome-Utica SMSA					
Herkimer	100	0.7	0	0	0.1
Oneida	100	70.8	71.8	12	0.5
Rochester SMSA					
Monroe.	100	84.1	89.3	39	3.3
Syracuse SMSA					
Madison	100	0	0	0	0.2
Onondaga	100	78.8	77.8	21	1.7
Oswego.	100	0	0	0	0.3

Sources: *U.S. Census of Population: 1960 -- General Population Characteristics, New York, Final Report No. 34B (Washington: U.S. Government Printing Office, 1961), p. 13.
**New York State Department of Audit Control.

Cleveland Area

Using the same criteria as those employed in determining urbanized areas in New York State, a comparable analysis of the Cleveland Metropolitan Area was undertaken. The restriction of the urbanized area to Cuyahoga County exclusively, partly because of common assessment throughout the county and partly because urbanization is still almost wholly within the county, shows that the measures used conform to the New York pattern. The urbanized area covers 98.6 per cent of the area and 99.9 per cent of the population. The density of valuation here again provides a measure of land use intensity which is of considerable value in analyzing local government problems in an urban context. Moreover, it permits the direct comparison of areas independent of the legal designation of the government, which changes arbitrarily in Ohio when a village reaches a population of five thousand either at Census time or when it has that number of voters.

The Cleveland area, with the exception of two small special cases, the villages of Bentleyville and Glenwillow, may be considered as a completely urbanized area, with differences in the degree of urbanization as indicated in Figures 2 and 4 below. A purist might eliminate the other parts of the outer ring, but two-to-four-acre zoning considerations which are operative in the eastern part of Cuyahoga County explain the relatively low valuations per square mile. However, using 42 per cent as the average assessment every municipality in Cuyahoga County would meet the $1 million-per-square-mile test used in New York State, except the two above-mentioned special cases.

SPATIAL AND LOCATIONAL PATTERN OF LOCAL EXPENDITURES

In many urban areas expenditure decisions are made by a variety of local governments, some of which are definitely overlapping. Insofar as the expenditure decisions are incomplete or are interrelated, any analysis must be expanded to include all the parts. We cannot look at a single government and generalize, nor can we look at a multiplicity of governments and assume that the multiplicity necessarily makes for higher expenditures. Preliminary analysis of New York State indicates that governmental complexity does not make for higher per capita municipal expenditures. Since the most complex units are primarily in the urban areas immediately outside the city boundaries, these become exceptionally important. Expenditures, at least measured in terms of per capitas or per assessed valuation, are not greater in the more complex areas as measured by the number of governments per person, governments per unit of area, or governments per county, than they are elsewhere.

Given the definition of urban areas in terms of governmental units and given a knowledge of the functional responsibilities of these governments, what can be said of the spatial and locational pattern of local expenditures in urban areas? How do spatial and locational measures supplement existing measures? To what extent do measures such as expenditures per capita of per $1,000 equalized valuation reflect the urban government decision-making process?

The preoccupation with per capita measures is the result of an interest in the questions of burden and benefit. While it is doubtless true that the per capita measure and its cousin the per student measure are related in most instances to the decision-making process, it seems that in the case of some <u>municipal</u> expenditures they are quite unrelated. The per capita values have to be explained, but they do not directly indicate the factors that are actually involved in the decision-making process. For the purpose of illustrating the problem let us look at three classes of functions in which expenditures per unit area appear to be of much greater importance than the per capita expenditures, because they are related to the decision-making process and in fact explain the per capita values. The three categories of functions are the police, the fire, and the total municipal operating function taken as a unit.[13] The areas chosen for analysis are the Cleveland area and the Rochester and Syracuse areas. As noted earlier, the former is a major urbanized area characterized by a very large built-up area involving many municipal governments; the latter two are smaller urbanized areas in which only a few local governments can be rightfully considered as wholly urban in nature.[14]

---

[13]Welfare expenditures are a county function in each instance and hence are excluded from the local municipal totals.

[14]Special districts are excluded from specific consideration but they are included under their appropriate town governments.

Police Expenditures

Variations in per capita police expenditures in the Cleveland area can be "explained" to a large measure by variations in per capita assessed valuation and to a lesser extent by variations in personal wealth as measured by the Ohio tax on intangibles.[15] When exceptions occur they involve small governments which have either extraordinarily high per capita assessed valuation or high per capita wealth. The spatial pattern of per capita police expenditures is, however, unrelated to the degree of urbanization. Only in the case of the slightly higher value of per capita police expenditures in Cleveland City proper relative to other cities can there be said to be any domination by the most urbanized area. However, if villages are included, the Cleveland value seems low by comparison to the industrial enclaves (urbanized, but not the same degree as Cleveland City) and to the wealthy suburbs (which although urbanized, do not approach that of Cleveland City).

What in fact is involved in the public expenditure decision-making process in the urban community? Particularly of interest are the questions of (1) what effect does urbanization have on expenditures, and (2) what are the effects of the large agglomeration of people and property on such things as police, fire, highways, and other expenditure decisions?

In dealing with spatial and locational aspects of local government expenditures, there is a question as to whether the expenditures per unit area might not provide the clue to the decision-making process of many classes of urban government expenditures. If there were a pattern related to urbanization it might show up in this type of analysis. It is quite obvious that expenditures per unit area will vary, but will they follow any pattern related to urbanization? Per capita figures are not too encouraging, but in combination with density figures, as expenditures per unit area, they are directly related to the measure of urbanization.

Police expenditures per square mile in the Cleveland area do not show the same pattern as they do in the case of per capita values. As shown in Figure 2, except for townships, there is no clear dividing line between cities and villages. Spatial and to a lesser extent locational considerations dominate the pattern. Moreover, they show a pattern approximating that of the valuation densitites which provide the criterion of urbanization. Expenditures for police protection per square mile decline in all directions as one moves outward from the central core of the area regardless of the expenditures per capita and per $1,000 of "equalized" valuation. Specifically, police expenditures range from $145,000 per square mile to $2,000 per square mile in the periphery.

Further, the police expenditure per square mile shows similarity in contiguous communities in the periphery where per capita measure indicates great diversity. (See Table 2.) As shown in Figure 2 there are no real exceptions to the spatial pattern, although there are some items which appear slightly out of line in terms of the measures of urbanization. These minor exceptions are of two kinds: (1) either they reflect an understatement of the amount of real property in a community or (2) they reflect the fact that the jurisdiction involved plays an important role with respect to

---

[15]Seymour Sacks and William F. Hellmuth, Jr., Financing Government in a Metropolitan Area (New York: The Free Press of Glencoe, Inc., 1961), p. 121.

traffic flows. In the former case the degree of urbanization is understated because of the failure to include some kinds of property, which under Ohio law are classified as tangible personal property, but which in other states are considered as real property (e.g., boiler plants, oil storage plants, and blast furnaces of steel mills).[16] The extent to which police expenditures in Cleveland City, Cuyahoga Heights village and Brook Park village (now a city) depart from the pattern may be accounted for by these exclusions from the measure of urbanization.

The other class of community which shows a departure from the localization pattern, but not the spatial pattern, is the community which performs the function of a traffic conduit and which because of its relatively small size involves very high police expenditures per square mile. This class includes Bratenahl, Woodmere, and the City of East Cleveland. In the former two villages extensive recourse is made to fines as a means of financing the very large police expenditures per square mile.

Table 2. Alternative measures of police expenditure in the Cleveland area, 1957*

Governmental Unit Cuyahoga County	Per capita	Per $1000 assessed valuation (equalized)	Per cent of budget	Per square mile ($ thousand)	No. of full-time police per sq. mile	Full valuation** per square mile ($ million)
Cities						
Bay Village.	$ 5.30	$ 1.90	14.3	$ 14	2.2	$15.7
Bedford	8.00	2.90	13.1	21	3.3	14.3
Berea	7.20	3.30	15.3	22	3.6	13.7
Brooklyn	8.40	1.80	20.8	19	2.9	14.3
Cleveland.	11.80	4.10	19.9	145	25.6	61.7
Cleveland Heights . .	7.30	1.80	16.2	55	7.9	48.6
East Cleveland	9.50	2.30	27.2	125	19.7	65.0
Euclid	6.60	2.10	20.0	37	5.9	30.0
Fairview Park.	7.10	2.70	17.9	27	4.1	23.8
Garfield Heights . . .	4.00	2.20	20.7	20	4.3	19.0
Lakewood.	6.40	2.50	9.4	76	11.3	60.2
Lyndhurst.	5.20	2.70	20.1	16	2.4	16.2
Maple Heights.	5.50	2.50	22.3	28	4.3	22.1
Mayfield Heights . . .	6.40	2.70	21.9	17	2.9	13.3
North Olmsted.	5.40	2.30	16.4	6	1.0	5.5
Parma	5.00	2.00	20.3	16	1.8	15.3
Rocky River	7.40	2.20	14.2	25	3.8	25.0
Shaker Heights	10.60	2.50	18.2	57	8.1	52.4
South Euclid	6.70	2.60	19.9	37	5.1	31.0
University Heights . .	8.50	2.70	23.9	74	13.3	62.9
Villages						
Beachwood	11.40	2.30	27.2	12	1.8	10.5
Bedford Heights. . . .	6.00	1.40	27.9	6	1.5	4.8
Bentleyville	25.10	7.70	58.7	3	.4	.7
Bratenahl	55.50	13.20	45.5	66	12.4	11.9
Brecksville.	9.00	2.10	28.5	2	.4	2.1
Broadview Heights . .	6.50	2.50	33.0	2	.3	1.9
Brooklyn Heights . . .	7.80	2.70	23.4	6	.5	4.5
Brook Park.	42.60	1.40	22.1	24	2.1	8.6
Chagrin Falls	12.20	3.70	26.0	20	2.9	9.5
Cuyahoga Heights. . .	117.60	1.00	22.3	63	5.1	29.0
Gates Mills.	30.70	4.50	29.4	4	.8	2.1
Glenwillow	1.90	.40	11.3	x	0	1.0
Highland Heights . . .	5.60	2.10	28.0	3	.2	2.4

- continued

[16] Ibid., p. 223.

Table 2. Alternative measures of police expenditure in the Cleveland area, 1957*, Continued

Governmental Unit Cuyahoga County	Per capita	Per $1000 assessed valuation (equalized)	Per cent of budget	Per square mile ($ thousand)	No. of full-time police per sq. mile	Full valuation** per square mile ($ million)
Hunting Valley.	69.40	6.20	40.9	4	.9	1.7
Independence.	8.60	2.60	29.1	6	.8	4.8
Linndale.	2.80	1.60	11.6	11	0	14.0
Mayfield.	11.70	4.50	40.4	5	.5	2.4
Middleburg Heights .	6.60	1.80	21.5	5	.9	4.8
Moreland Hills	16.60	4.10	40.8	4	.7	2.1
Newburgh Heights . .	7.90	3.20	29.2	52	7.3	32.1
North Randall	74.70	4.10	40.9	34	0	14.3
North Royalton	7.70	3.00	34.5	2	.4	1.7
Oakwood.	3.50	2.70	25.8	3	.3	2.9
Olmsted Falls	7.20	2.90	23.6	9	7.0	6.9
Orange.	12.70	4.00	56.1	6	1.2	3.6
Parkview	6.60	2.60	34.7	11	0	9.8
Parma Heights	7.30	3.10	23.5	4	3.0	16.0
Pepper Pike	16.10	2.40	29.1	6	.8	4.5
Richmond Heights . .	7.20	3.10	43.5	6	.9	4.0
Seven Hills	9.20	3.20	32.7	7	.8	4.5
Solon	8.10	2.10	27.0	2	.3	2.1
Strongsville	10.10	3.80	31.7	2	.3	1.1
Valley View	15.50	5.00	38.8	3	.5	1.1
Walton Hills	14.50	.60	44.0	3	.4	5.5
Warrensville Heights	6.40	2.60	22.7	13	2.3	11.4
Westlake	6.30	2.20	28.6	4	.9	4.0
West View	6.20	2.50	27.5	3	0	2.9
Woodmere	25.40	12.00	54.2	27	2.5	5.5
Townships						
Chagrin Falls	0	0	0	0	0	2.1
Olmsted.	1.70	1.00	30.3	1	2.0	1.4
Riveredge.	0	0	0	0	0	n.c.
Warrensville.	.50	0	23.8	x	0	5.5

Sources: *Protection in Cuyahoga County, Cleveland Metropolitan Service Commission, 1958, pp. 60-63.
**Financing Government in a Metropolitan Area, pp. 30-32. Using a 42% equalization rate.

n.c. - Not computed.
x Less than $500 per square mile.

The regularity of the spatial pattern is only partly the result of population densities. Relative differences in population density do not explain the similarities of expenditures per square mile in the periphery, nor do they explain similarities in the inner area including the core. The total urbanization mix, people and/or property, does explain the pattern. As shown in Table 2 the expenditure indicates a range of protection from twenty-six police per square mile to less than one policeman per square mile in the periphery.

The differences in per capita expenditures appear to be explained by the relative amounts of protection per square mile. The differences between Hunting Valley's $69.40 per capita, Gates Mills $30.70 per capita, and Moreland Hills $16.60 per capita would seem to arise in part from providing the approximately same level of expenditure ($4,000) per square mile. The decision making involved the provision of certain levels of expenditures for police per unit area. There is some doubt that the local officials are even aware of the per capita figures.

The same is true of the industrial enclaves and some of the inner ring cities. What does the $117.60 per capita expenditure for police in Cuyahoga

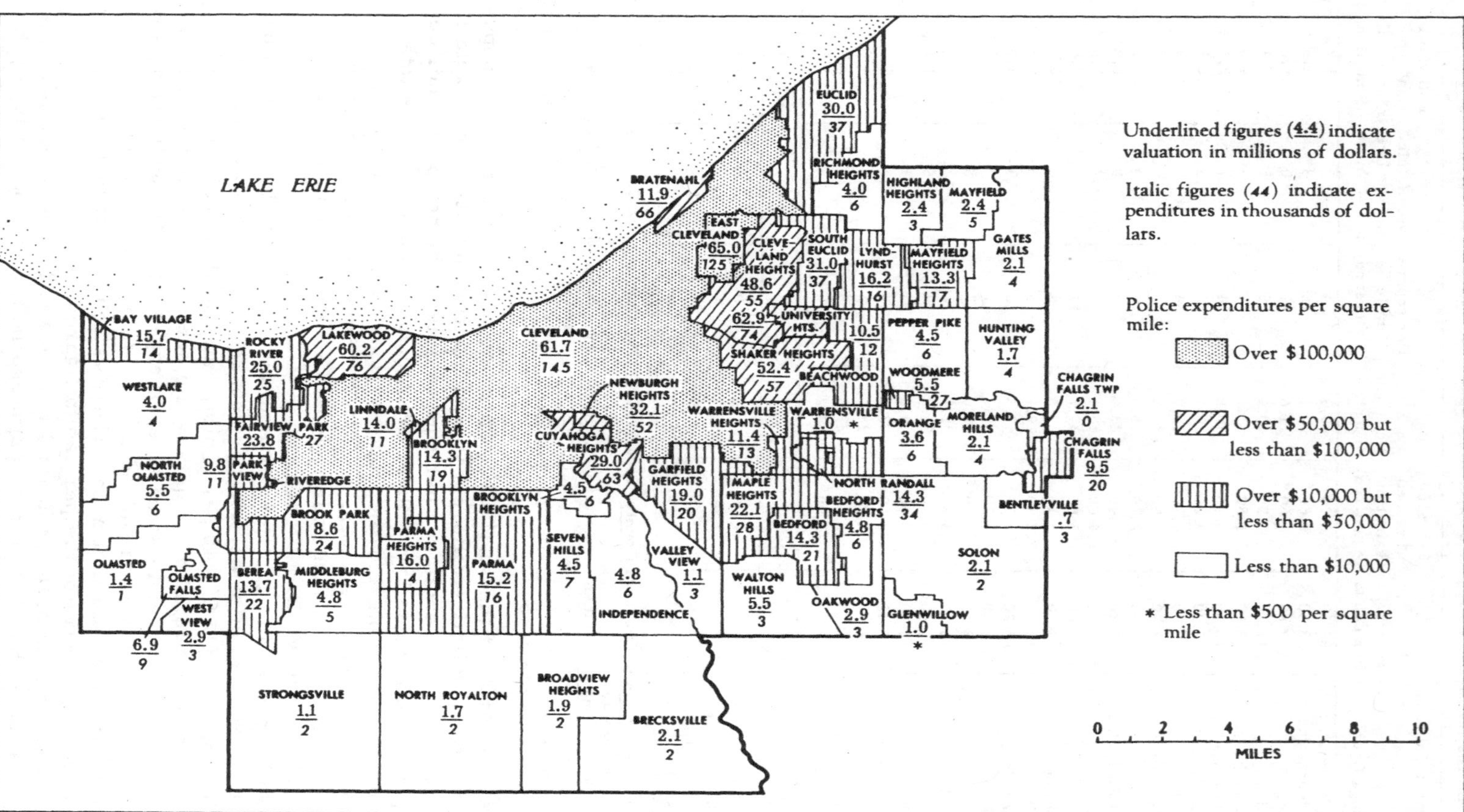

Figure 2. Police expenditure per square mile and estimated full valuation of real estate, Cleveland metropolitan area, 1956.

Heights mean? What does the $42.60 per capita expenditure in Brook Park mean? Placed in terms of expenditures per square mile, they involve the same police expenditures as provided by Cleveland Heights, and Parma Heights expenditures of $7.30 per capita. Are they not all providing a certain amount of protection per unit area which reflects the degree of urbanization?

If it is argued that police expenditure per square mile is a function primarily of the degree of urbanization, then the per capita costs will vary according to the number of people relative to the degree of urbanization. The high expenditures per capita in the wealthy suburbs, industrial enclaves, and central city all reflect a response to their relative urbanization, and not their population density. These very high per capita expenditures mean something quite different when interpreted as expenditures per square mile. They usually involve modest expenditures per unit of area.

Correlation analysis confirms the existence of a strong positive relationship between police expenditures per square mile and the amount of property per square mile. The simple coefficient of correlation between the two variables is +.96. The regression of property per square mile on police expenditure indicates that for every additional million dollars of real property per square mile there is an additional police cost per square mile of $1,624.

When applied to the Syracuse and Rochester areas where the density of valuations indicate that the urban areas are much more limited in size, the general pattern of expenditures found in the Cleveland area is, nevertheless, duplicated. In the Syracuse area, for instance, only the city of Syracuse and three towns have police protection in excess of $1,000 per square mile. Of course, individual villages achieve higher amounts but they involve areas in no instance larger than two square miles, as opposed to the much larger dimensions in the Cleveland area. Only in one town in the Syracuse area, that of Geddes, with the highly industrialized village of Solvay does the police expenditure per square mile ($8,000) appear appreciable.[17] Even if the town police expenditure in each case is just for the town outside the village then only in Geddes will this appear as any but a nominal amount. Some police protection is provided by Onondaga county, but here again the protection is not of an urban nature.

In the Rochester area the picture falls between that of Cleveland and Syracuse, but much closer to the latter. Again there are no exceptions, as expenditures for police move with urbanization. Police expenditures per square mile show the same pattern as urbanization. In the Rochester area, however, there are two highly urbanized towns, Brighton and Irondequoit, with police expenditures as high as $14,000 and $10,000 per square mile respectively.[18]

Fire Expenditures

This spatial phenomenon is related not only to police, but to fire, streets and highways, and the total of current municipal expenditures. (It should be noted again that welfare expenditures are not included because they are provided by the county governments in the Cleveland, Syracuse,

---

[17] Comparison of Revenues, Expenditures, and Debt, op. cit. (footnote 6), p. 104.

[18] Ibid., pp. 90-93.

and Rochester areas.) In the case of fire a number of important provisions must be considered at the outset. First, if the presence of full-time protection is made a criterion of urbanization, then regardless of the measure, the urban areas would be limited. The inclusion of this function has very distorting effects on what may be called the common functions, except in the very largest of cities. Second, since agreements between governments and the purchase of fire service by one government from another are common in urban areas, the expenditures, no matter how measured, may be severely limited. Finally, there are strange admixtures of full-time and volunteer fire companies in urban areas whose dollar expenditures per square mile in no way represent the true amount of protection. If account is taken of these three factors, as is done in Figure 3, then the spatial pattern indicated in the analysis of police expenditures appears quite clearly.

In Cleveland and in most other places the boundaries of full-time fire expenditures are considerably narrower than the urban area as indicated by the density of valuation. Perhaps this influences total expenditure to a greater extent than is usually assumed in analysis of urban expenditures. While the city of Cleveland dominates fire expenditures per square mile (although not to the same extent that it does in total expenditures) it is followed by Lakewood and East Cleveland whose other characteristics link them to Cleveland in an inner core area. By a like token Cleveland Heights, University Heights, Shaker Heights, and Cuyahoga Heights show a similarity in fire expenditures per square mile which is roughly proportional to their density of valuations.[19] These latter show great variations in per capita expenditure, but similarity in the amount of protection they provide per unit area. Even similarities in paid firemen-volunteer firemen ratio appear as a spatial phenomenon.

It is of exceptional interest that fire protection was not explained by multiple regression analysis, using the standard measures by class of government. The pattern appears in large measure to be urban and spatial, independent of the class of government.

A major implication of this analysis is that a community cannot afford the luxury of <u>not</u> having a full-time fire department after a certain degree of urbanization has been reached. This of course is not true as a matter of fact, but it might become increasingly true in terms of the kinds and amounts of special arrangements where the actual levels of protection go beyond the boundaries indicated.

All Municipal Expenditures

The last step in this experiment was to test whether the expenditure patterns go beyond those of the individual function to the total of all municipal functions--to test the extent to which the pattern is altered as different governments assume alternative responsibilities. In the Cleveland area, the city, village, and town governments do not bear responsibility for schools, libraries, and welfare. Hospital and utility expenditures are excluded because of the nature of their financing.

[19]In the case of Cuyahoga Heights real estate, full valuation was understated because of the exclusion of blast furnaces from the real estate base.

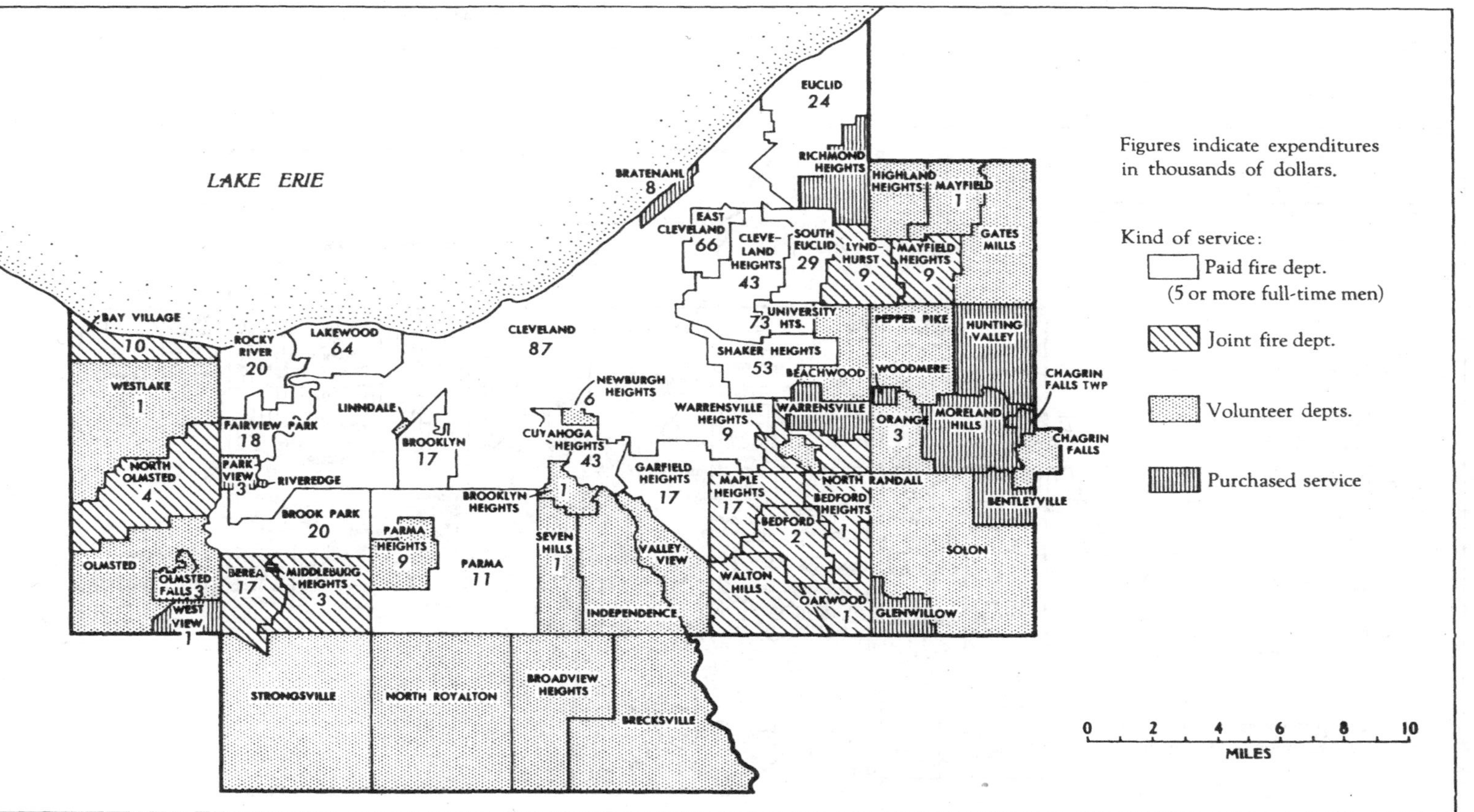

Figure 3. Fire protection services and fire expenditures per square mile, Cleveland metropolitan area, 1956.
Source: Fire Protection in Metropolitan Cleveland, 1958, pp. 6-8.

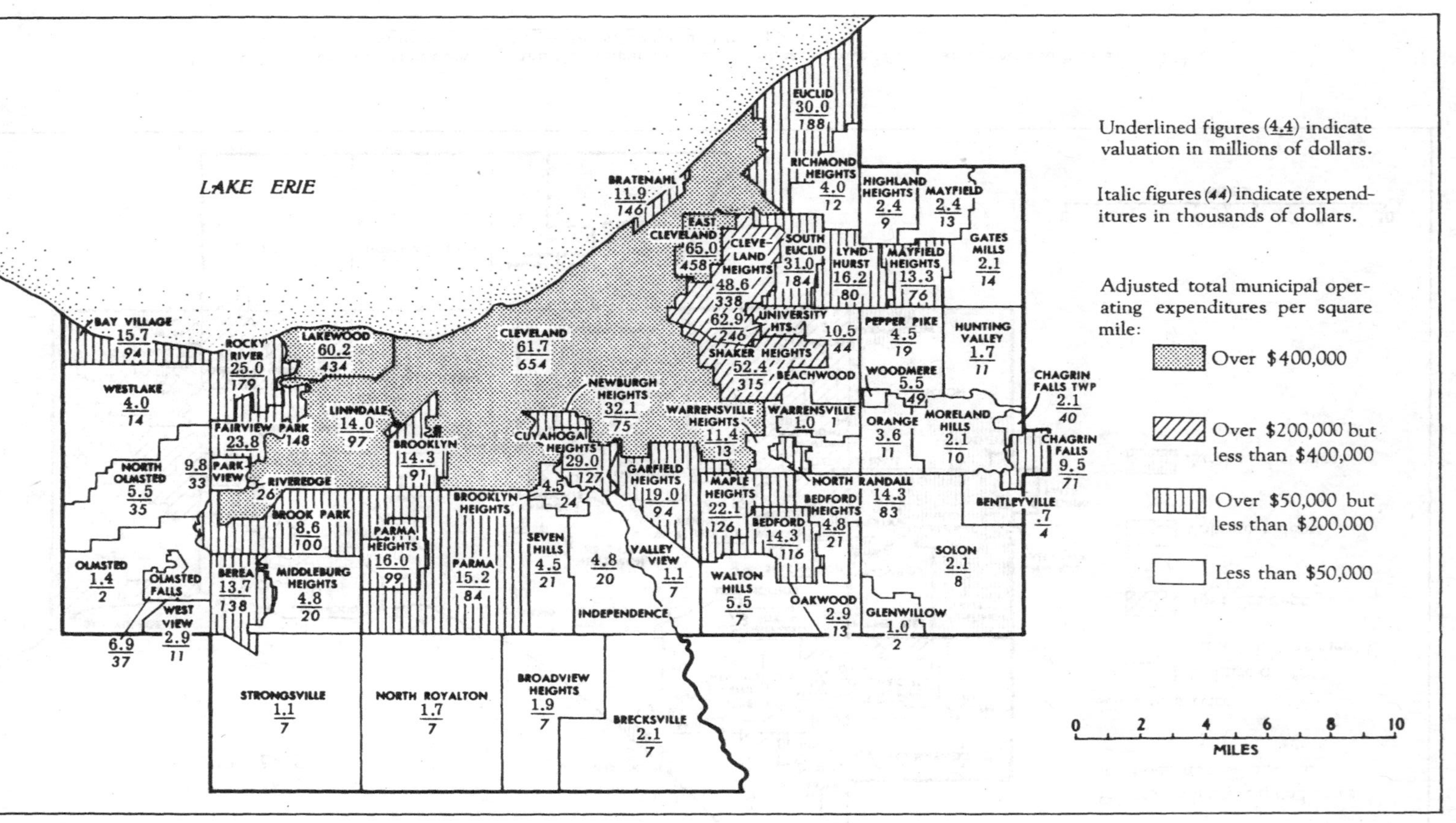

Figure 4. Estimated full valuation of real estate and total municipal operating expenditures (excluding hospital expenditures) per square mile, Cleveland area, 1956.

Source: Financing Government in a Metropolitan Area.

The spatial pattern that emerges in analyzing police and fire expenditures also appears when all municipal expenditures are considered. Once again industrial enclaves and wealthy suburban communities appear to be reflecting their degree of urbanization more than any other individual factor, as shown in Figure 4. Total municipal expenditures per square mile are highest in the central city even without the inclusion of welfare expenditures and move downward in all directions. The amounts again appear to be independent of the per capita expenditures. The pattern of expenditures per square mile places each degree of urbanization in its own spatial grouping. Exceptions occur as communities purchase or are given services by other governments, but the over-all pattern remains the same. As in the case of police and fire, correlation analysis shows a striking association between the property value per square mile and the total expenditures per square mile ($r = +.92$). The regression coefficient indicates that for every additional million dollars of valuation per square mile there are total municipal expenditures per square mile of $6,711.

The striking difference in urbanization in the Syracuse and Rochester areas as compared to the Cleveland area is reflected in the expenditures per square mile. The levels simply do not approach those of the Cleveland area, and these are not due to the assumption of responsibilities by the county.

CONCLUSIONS

1. Public finance data provide a useful way of classifying local governments in an urban-nonurban continuum. Valuation densitites per square mile provide a quantitative measure of the degree of urbanization which is independent of the legal classification of the government. This is not only of value in providing a framework for analyzing local government decision making in urban areas, but this provides a quantitative measure of urbanization in terms of governmental units which may be used in other areas of analysis as well.

2. There are two classes of urban areas--a very limited number of large urban areas, and a larger number of small urban areas. In the former case the effects of urbanism are reflected in the levels of governmental services provided by numerous governments, in the latter case the number of governments providing urban levels of service appears to be relatively small. Population size or density is not necessarily a measure of urbanization in analyzing individual governments.

3. In order to analyze individual local government decision making in urban areas it is essential to know the pattern of state aid and the allocation of responsibilities by local governments. Where state aid is large relative to total expenditures, the crucial questions before the local government may involve tax decisions rather than expenditure decisions. In the case of municipal functions in urban areas, exclusive of welfare and to a lesser extent highways, the decision-making process is perhaps closest to being truly local.

4. While expenditures per capita and per $1,000 equalized full valuation do not show any spatial or locational pattern or show any pattern related to urbanization, various classes of municipal expenditures per square mile do show a definite spatial and to a lesser extent locational pattern which directly reflects the extent of urbanization as measured by the

density of valuation. This appears to be the case of police, fire, and total current expenditures. Because of other factors, there are some classes of expenditure which are only in a general sense related to the extent of the urbanization of a political jurisdiction (e.g., education and welfare). Different levels of expenditures are associated with different densities of valuation. Nonurban levels of expenditures are associated with nonurban areas, regardless of the per capita expenditures involved.

5. The levels of expenditures as measured in per square mile terms may in turn help explain the per capita variations, not only in the amount of the variation but in the spatial and locational characteristics of the variation. Thus, in the Cleveland area there are two industrial enclaves with entirely different patterns of expenditures. However, one involves intensive highly urbanized land use and the second extensive, less urbanized, use of the land. The per capita amounts seem to be reflected by the fact that different levels of services are needed in these two areas, and given their populations and resources, these are then determined. This seems to be true of other kinds of communities as well.

6. It now seems quite clear that while per capita measures do not show any regularities and, hence, may not provide a good method for projecting municipal expenditures in urban areas, the expenditures per square mile may in fact do so. This should not preclude the use of per capita measures for comparative purposes. The recognition that expenditures may be property-oriented, person-oriented, or the result of special conditions will enhance our understanding of the link between urbanism and governmental expenditures.